INVENTORY 1985

BATTLE FOR SPACE

BATTLE FOR SPACE

Curtis Peebles

CONSULTANT **Kenneth Gatland** FRAS FBIS

Beaufort Books Inc.
New York   Toronto

Acknowledgements

This book would not have been possible without the help and advice of many people. First, and most importantly: John Davies, Martin Mulligan, Kenneth Gatland and everyone else at Grub Street for turning my manuscript into a book. Also helping were: Dayn T Beam, Henry E Catto Jr, Peter Campbell, George Franklin Jr, Phil Henderson, Charles W Hinkle, Milton Jordan, Cpt Barbara J Koll, Robert J McCormick, Harvey J Nathan, Guy B Oldaker, Joseph Olivo Jr, K B Patton, Col James H Rix, Shirley Sontag, Capt Billy J Stokes, F P Thompson, William V Vitale and Anne Wilkinson.

Curtis Peebles,
El Cajon,
California,
1983.

Library of Congress Cataloging in Publication Data
Peebles, Curtis.
  Battle For Space.
  1. Astronautics, Military. 1. Title.
UG1520.P43 1983   358'.8   83-9930
ISBN 0-8253-0160-2

Published in the United States by Beaufort Books, Inc., New York.
Published simultaneously in Canada by General Publishing Co. Limited.

Designed by Grub Street.

Printed in Great Britain.   First Edition
10   9   8   7   6   5   4   3   2   1

Contents

Foreword

The individual seeking information on the space weapons race faces many difficulties. Military space missions are wrapped in an all-encompassing blanket of secrecy, both reasonable and unreasonable. The most famous example concerns reconnaissance satellites. The Kennedy Administration limited very strictly the amount of information released about reconnaissance satellites in order to protect US intelligence-gathering sources, and would neither officially confirm nor deny it actually used them. Such secrecy even came to encompass military weather and navigation satellites, and persisted for the better part of two decades.

In dealing with Soviet space activities, the analyst faces an even more daunting problem. To this day, the Soviet Union flatly denies it undertakes *any* military missions and angrily lambasts the US for militarizing space. In reality, the Soviets maintain a military space effort far in excess of the US equivalent. Whenever information about the Soviet space program is acquired, it is classified by Western intelligence agencies.

As a result, information tends to be fragmentary and not readily accessible, stumbled across intermittently in obscure or specialized publications. These may include analyses of US and Soviet strategy and such items as reports on the statements to Congressional budget committees, along with official and unofficial leaks. In trying to determine the mission of a reported Soviet space launch, the analyst can however use its published orbital characteristics – the height of its orbit, the inclination and the launch vehicle. Another indicator is orbital behavior – does the satellite maneuver; how long does it stay in orbit; was it recovered after re-entry? Answers to these questions provide clues.

This book also uses documents which were formerly secret but have now been released under the Freedom of Information Act. This Act allows an individual to request the declassification of government documents; the documents in question are then reviewed and those portions currently and properly classified in the interest of the national defense of the United States are deleted. Nevertheless, the Act can provide a wealth of data. In *Battle For Space* the sections which describe Program 437 and SPIKE owe their existence to the Act. Information previously published about these programs was minimal and misleading. Here is the full story.

Numerous articles and books, along with hundreds of pages of documents, were sought out in the course of the book. Only the most reliable and up-to-date information has been used. Speculation has been intentionally limited and conjecture, where it appears, is informed.

References have been given in order that material may be checked for reliability and further information sought.

INTRODUCING THE BATTLE FOR SPACE

Kenneth Gatland FRAS FBIS

'We are not reaching for prestige in space – we are reaching for peace (...) We seek to make space an instrument for peace and the development of mankind. But if we abandon the field, space can be pre-empted by others as an instrument for aggression.'

Vice President
Lyndon B Johnson,
speaking at the
University of Maryland,
June 8, 1963

In a top security building in Washington, United States intelligence experts peer at a color photograph enlarged to revealing detail on a viewing screen. They are looking at an airfield deep inside the Soviet Union where a new Soviet bomber sits on a runway.

The photograph is one of a huge number taken by United States Air Force reconnaissance satellites which are bringing to light, in astonishing detail, many of the most closely guarded secrets. Missile sites, airfields, cosmodromes, factories, shipyards, docks, Warsaw Pact maneuvers, world shipping, Communist infiltration overseas — all such manifestations of the Soviet war machine are yielding to the uncompromising stare of the USAF's Big Bird and KH-11 satellites on routine patrol. On some photographs, obtained under ideal atmospheric conditions, it is possible to see people walking the streets of towns, and even to read the identification numbers of aircraft on airfields.

The Soviets could undoubtedly show similar secrets of the United States, Britain and other NATO countries gleaned from their own Cosmos space spies.

It has to be admitted that spying from space has become a way of life because, in an imperfect world, the United States and the USSR distrust each other's intentions and feel the need to police international agreements on arms control.

The late President Johnson once remarked that United States reconnaissance satellites saved the cost of the space program many times over because the information they provided prevented over-reaction to uncertain threats and actually limited military spending.

Right: A right turn off the Colorado Highway 115, Cheyenne Mountain is a hollowed-out nuclear refuge in the Rockies, housing a 4.5-acre complex of computers, communications systems, display screens and technicians. As the home of NORAD, it is the place from which any aerospace battle over the US would be coordinated and has hot-line links to the Pentagon and White House. Blast doors weighing 25 tons each and over 3 ft (0.9 m) thick protect the community of 900 people from nuclear attack. All the buildings rest on steel springs which act as shock absorbers. Blast and heat from an explosion would be channelled through the tunnel and vented on the other side of the mountain. Incoming air is stringently filtered against chemical, biological and radioactive agents, and protective measures against EMP are also included. The vast quantity of granite removed when excavating the mountain was piled up to form the car park (inset).
Below: Norad's crest.

The frequency with which such satellites are launched, and their ground tracks, reveal that special efforts are made to observe military conflicts around the world. A good example was the Arab-Israeli war of 1973 when the Soviets put up six spy satellites in a period of 21 days. In similar vein they inspected the Iran-Iraqi war and the battle for the Falklands.

The superpowers also keep watch on what is happening in space. Almost from the day the West was rocked to its foundations by the launch of Sputnik 1 on October 4, 1957, the United States Air Force has maintained close scrutiny of each new satellite which appears in orbit. The information, which comes in a steady stream from a network of advanced radars and other satellite-watching devices in various parts of the world, is immediately processed by computers for fast recall.

The place where this intelligence comes together has a distinctly Dr Strangelove flavor. The Air Force personnel monitoring this activity, forming part of North American Aerospace Defense Command (NORAD), are housed in a steel building within a chamber carved from the solid granite core of Cheyenne Mountain near Colorado Springs. If ever the world goes mad and nuclear-tipped missiles fly like arrows between continents, this could be one of the last outposts of civilization.

On September 1, 1982, the United States established within this complex a Space Command whose duties include Space Shuttle military operations and control of two F-15 Eagle squadrons which will launch the USAF's new anti-satellite missile.

This is partly in response to the Soviet development, within the Cosmos program, of 'hunter-killer' satellites which have been extensively tested against orbiting targets. Their aim is to track down 'enemy' spacecraft orbiting close to the Earth and disable them with shrapnel during a flypast.

The US space defense element has five basic objectives:
- To detect, track and identify all objects in space;
- To provide warning of hostile acts in space and provide protection to aerospace resources;
- To provide timely warning to the national command authorities of hostile actions against the United States and its allies;
- To enhance deterrence by developing the capability to deny or nullify hostile acts in or through aerospace;
- To 'be capable of conducting sustained operations to detect and analyse aerospace threats.'

Space war — the moral dimension

In 1983 the US Secretary of State, Caspar Weinberger, pointed out that some 85 per cent of Soviet space efforts were for military or combined military/civil purposes.[1] The Soviets launch nearly 100 spacecraft a year,

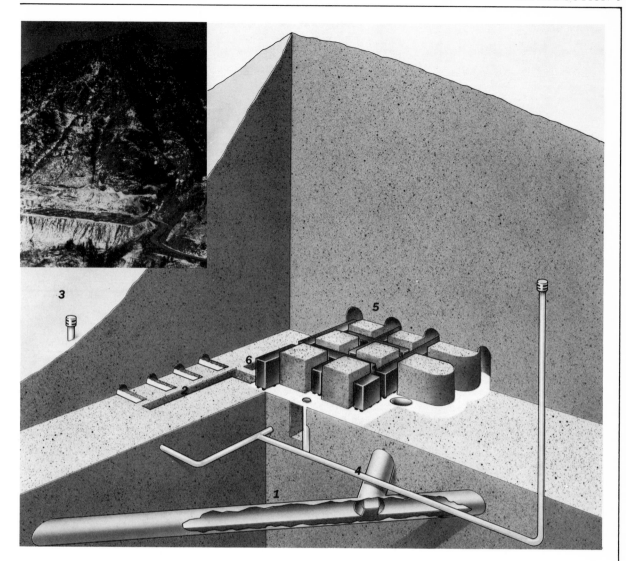

of which some 50 per cent are reconnaissance satellites; and other spacecraft are in orbit for early warning, navigation and communications purposes. While the United States in general gets more value from its fewer satellites, the annual payload weight placed in orbit by the USSR — ten times that of the United States — indicates a rapidly expanding capability in civil and military terms.

Against this, Soviet military commentators point to greater US emphasis on military space ventures in the age of the Space Shuttle. Of 70 major spaceflights planned by the United States through 1987, 25 are to be fully devoted to 'extending the Pentagon's secret assignments.'

Few observers of the space scene in the United States would deny this. By 1983, there were certainly more people in military uniforms at Cape Canaveral than in the 'purist' days of the Apollo Moon program, while preparations to launch

1 Central tunnel
2 Reservoirs – storing 6 million gallons of water from Colorado Springs
3 Air intake vents
4 Steel blast doors
5 Command post
6 Steel shells supported by steel frames

Space Shuttles from Vandenberg Air Force Base on the west coast reinforce the view that space is becoming the new military arena. On the other hand, there has always been a military presence at Soviet cosmodromes, since all launchings are carried out by the Strategic Rocket Forces.

Is it better to fight above the Earth rather than on it? The plain fact is that there is no longer a demarcation line between targets in space and on the Earth below. Satellites are a vital arm of military technology. Their functions include vastly improved communications between military forces and command centers; more precise navigation of ships, aircraft and missiles; improved reconnaissance and surveillance of other countries, including the movement of ships and aircraft, and more rapid and reliable advance warning of attack.

US Air Force General Tom Stafford (ret), the former astronaut, once told a Senate Committee: '(. . .) space may be viewed as an attractive area for a show of force. Conflict in space does not violate national boundaries, does not kill people and can provide a very visible show of determination at relatively modest cost.'[2] But if one country began destroying defense satellites, the trip wire of nuclear war would be close to snapping.

Nor is it war activity alone which worries the Soviets. The security of the Soviet state is based on rigid control of the press and broadcasting media. If the United States or anyone else used direct-broadcasting satellites to beam subversive material directly into the USSR or any other member of the Eastern bloc, they would reserve the right to destroy them. Presumably this would include the broadcasting of all kinds of propaganda, pornography and even religious material.

We are only now entering the field of direct broadcasting of television in Western Europe which, without some restriction, could easily overlap into Warsaw Pact countries.

Although it was the Soviet Union that first embarked on orbital weapons — the Fractional Orbit Bombardment System (FOBS) abandoned in 1980, and the 'hunter-killer' satellites now operational — it was already obvious, when talks seeking to curb the development of anti-satellite weapons opened in Helsinki in June 1978, that Soviet delegates were far from happy. They particularly disliked the re-usable Space Shuttle the United States was developing to reduce the cost of satellite launching, as they were afraid the US Air Force might use it to intercept their satellites. In theory the Shuttle could pluck a satellite out of orbit and fly it back to the United States. The spaceplane's cargo bay — 60 ft (18.3 m) long and 15 ft (4.6 m) wide — is large enough to accommodate some of the biggest man-made objects in space.

They were also concerned that the United States might use the spaceplane as an orbiting weapons platform for launching infrared homing missiles at their satellites, or even develop it as a laser gunship.

Despite protests from the Americans that nothing of the kind was contemplated — and in any case an anti-satellite treaty would have built into it the principle of non-interference with another nation's spacecraft — the Shuttle remained a bone of contention.

Although the talks continued in Geneva and Vienna in 1979, the Soviet invasion of Afghanistan caused Washington to soft-pedal.

The Kremlin makes a move

It was the Kremlin that took the initiative to get negotiations re-started. In 1981 Mr Andrei Gromyko, the Soviet Foreign Minister, made a direct appeal to the United Nations' General Secretary for a treaty banning the deployment of any weapon in outer space. The Soviet Union, he said, 'was in favor of keeping outer space clean and free from any weapons for all time and of preventing it becoming a source of worsening relations.'[3]

A previous treaty banning weapons of mass destruction from space, signed by 69 nations in 1967, led to the Soviets dropping their own project for an orbital bomb (FOBS); but the very existence of America's Space Shuttle made them extremely reluctant to abandon their own satellite defense system.

By mid-summer 1982, the Communist newspaper *Pravda* was claiming that 'Washington's plans, aimed at militarizing outer space, considerably heightened the threat of a nuclear conflict breaking out.'[4] The United States was accused of blocking the drafting at the United Nations Disarmament Committee of an agreement banning the placing of weapons of any type in outer space. At the same time, at a meeting of the Political Consultative Committee, Warsaw Pact member states renewed their call for an immediate start to negotiations to ban the emplacement of weapons of any kind in space.

The call was repeated in the Soviet press (as it is constantly for nuclear disarmament) to a public kept totally ignorant of the fact that the USSR has developed space weapons and is doing everything it criticizes the West for doing in its own defense.

It was against this background that in March 1983, President Reagan made his surprise announcement of a 'Star Wars' umbrella of ray weapons to shield the United States from nuclear attack. This envisaged stationing beam weapons in orbit and appeared to be in breach of the SALT 1 treaty of 1972 aimed at keeping the destructive capability of both sides roughly in balance.

The relevant section reads: 'Each party undertakes not to develop, test or deploy ABM (anti-ballistic missile) systems or components which are sea-based, air-based, space-based or mobile land-based'.[5]

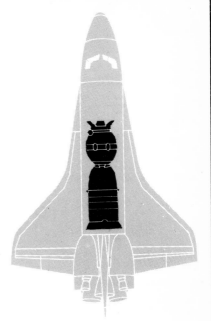

Soviet objections to the Space Shuttle center on the fact that its cargo bay can comfortably accommodate most satellites. It can theoretically pluck an enemy satellite out of orbit, as is happening here with the Soviet Soyuz craft.

NB On this and subsequent pages US hardwear is shown in grey, the USSR's in black.

Major Sites of Interest

Some of the key centers in the battle for space are shown on this map. They include places where ballistic missiles and military satellites are launched and tested. Also marked are the ground stations of competing powers which monitor them. Inside the US nerve center NORAD, the vital Space Defense Operations Center combines a computerized satellite-tracking network with command functions for military satellites, Space Shuttles and anti-satellites. Sites of equivalent importance in the USSR are the Baikonur and Northern Cosmodrome launch sites. Also featured on the map are the installations where the superpowers conduct research into beam weapons under the tightest security.

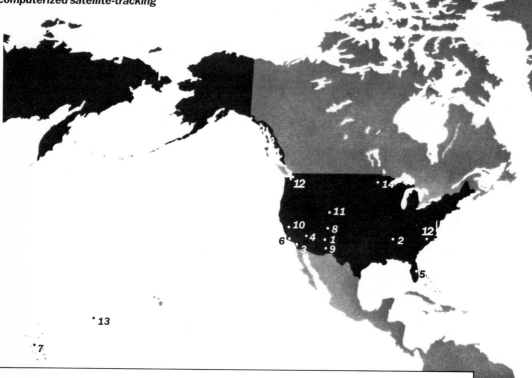

USA
1 USAF Weapons Laboratory, New Mexico, where a laser destroyed two pilotless aircraft, November 1973.
2 North Alabama, where in 1976 the US Army's Mobile Test Unit shot down aircraft and helicopters.
3 San Juan, Capistrano, California. In 1978 a US Navy laser destroyed anti-tank missiles in flight here.
4 China Lake, California. The USAF NKC–135 Airborne Laser Lab (ALL) began tests here in June 1981 against air-to-air missiles. The ALL is based at Kirtland AFB, New Mexico.
5 Kennedy Space Center, Florida. Space Shuttle operations began here in 1981.
6 Vandenberg Air Force Base, California, plans to operate Space Shuttle in 1985 and is an important ASAT-linked location.
7 Kwajalein Atoll, home of the earliest US ASAT, Program 505.
8 Los Alamos Laboratory, New Mexico, where the US DoD is spending $80 million on the White Horse particle beam project.
9 High Energy Laser National Test Range, White Sands, New Mexico, where the Sea-Lite laser has been tested.
10 Lawrence Livermore National Laboratory, near Oakland, California. Here the Chair Heritage project explores charged particle beams.
11 NORAD, in the heart of Cheyenne Mountain.
12 Two airfields from which the ASAT F-15 is intended to operate.
13 Johnston Island, home of the high-altitude ASAT, Program 437, before the devastating hurricane.
14 Project Safeguard in North Dakota, the abandoned US ABM site.

USSR

1 Troitski: the major laser development center. Believed to be concerned with both civil and military applications.

2 Semipalatinsk nuclear test site. First mentioned by Major General George J Keegan, former head USAF Intelligence, as the major center for research into missile-destructive high-energy particle beams.

3 Saryshagan test range, where explosive-driven generators produce hundreds of megavolts. Concerned with testing laser weapons and particle beams against flying targets.

4 Baikonur cosmodrome, site of new activity relating to huge multi-stage rocket with six to seven times the launch capability of the US Space Shuttle.

Believed to be connected with development of 12–20 person modular space station to be established mid–1980s. According to the US DoD: 'The Soviet goal of having continuously manned space stations may support both defensive and offensive weapons in space.'

5 The Northern Cosmodrome near Plesetsk, where Cosmos satellites originate as targets for the Soviet 'hunter killer' satellites launched from Baikonur.

6 Sarova, south of Gorki, a particle beam research center.

7 Kazakhstan recovery zone, where the USSR's reconnaissance film capsules fall to Earth, to be recovered by mobile teams.

8 Moscow, the only Soviet site which can be legitimately defended by anti-ballistic missiles under US-Soviet agreements.

9 The point in the Indian Ocean where the Soviet 'Kosmolyot' was recovered after 1¼ Earth orbits.

The United States' first efforts to test anti-missile defenses began with Nike-Zeus. This culminated in the celebrated $6 billion Safeguard system installed at Grand Forks Air Force Base, North Dakota, to protect Minuteman ICBMs installed there. Safeguard depended on two types of ABM: Spartan, which attacked at long range above the atmosphere, and the fast-acting Sprint, designed to catch missile re-entry vehicles which had eluded Spartan after they entered the atmosphere. The first was to achieve its 'kill' by a thermonuclear warhead, the second by a low-yield nuclear explosion.

The SALT 1 treaty allowed the US and the USSR two ABM complexes each, one to protect the capital city and the other to defend an ICBM complex. In practice each side installed only one.

The Grand Forks complex was declared operational on October 1, 1975, only to be closed down by Congressional decree the next day.

EMP — the unforeseen effects

One reason was recognition of the magnitude of the problem of seriously blunting a determined nuclear attack, particularly when ICBM's would discharge multiple warheads, decoys and all kinds of debris to confuse the defenses.

Another was the damage it would inflict on the US itself if ever it was used in anger. Critics of the system had argued that if Safeguard intercepted enemy missiles above and within the atmosphere, its nuclear explosions would have knocked out important communications systems and logic circuits across the entire US continent. This is because widespread electromagnetic pulse (EMP) is generated by transfer of energy from gamma rays to electrons when a nuclear weapon is exploded.

Recognition of the impact of EMP on defense systems led to two main shifts in technical policy: greater reliance on geosynchronous satellites for military communications, and a rejection of nuclear explosives as the instruments of space defense.

The first US anti-satellite weapons, both nuclear-armed, had been deployed in the area of the Pacific in the early 1960s. Nike-Zeus was adapted for this role on Kwajalein Atoll while Thor stood ready on Johnston Island.

When treaty obligations and EMP finally ended the enthusiasm for nuclear attack systems, new technology in the shape of much improved weapons guidance began to be applied.

Henceforth, missiles would rely on shrapnel effect, or steer themselves directly into their targets. Alternatively, there was the more distant possibility of using ray weapons.

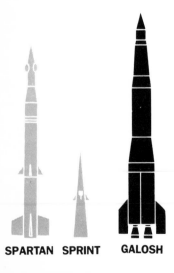

SPARTAN SPRINT GALOSH

The Soviet Galosh ABM in its launch cannister. It has a thermonuclear warhead and attacks at long range above the atmosphere.

All the same the Soviets installed their ABMs in the Moscow area, putting 16 missiles (NATO codename Galosh) into each of four sites with associated tracking and guidance radars. To give the Moscow defenders advance warning of a missile attack, they installed powerful perimeter radars at remote sites (eg, near Irkutsk in Siberia, in Latvia and near the Barents Sea).

The Moscow defense system can be regarded only as a token defense liable to be quickly overwhelmed by a determined assault by Hydra-headed ICBMs. The Galosh missile, which attacks at long range above the atmosphere, carries a thermonuclear warhead which bursts close to its target.

US satellite observation indicates that the Moscow ABM defenses have been modernized and that new, improved missiles have been deployed. At best they give some assurance against a limited attack from China, or a stray missile launched in error.

In his March 1983 speech to a nationwide TV audience President Reagan said the United States would move away from nuclear deterrence in favor of a new kind of ABM system — the beam weapon.

The Kremlin's response was predictable. Mr Yuri Andropov, the newly appointed Soviet leader, warned that any move by the United States towards setting up an antiballistic missile defense system involving laser beams and mirrors in space was 'extremely dangerous'. He regarded such proposals as being aimed at disarming the Soviet Union. 'All attempts to achieve military superiority over the USSR are futile,' he said. 'The Soviet Union will never allow them to succeed.'

As these issues are likely to have a major effect on future weapons policy, it is as well to look at the President's statement in greater depth. A major part of his speech had been devoted to convincing US taxpayers of the need to strengthen the United States' defenses so that it could negotiate arms reductions from a position of strength. 'There was a time,' he said, 'when we were able to offset superior Soviet numbers with higher quality. But today they are building weapons as sophisticated and modern as our own.'

Was there a way out of this dilemma? After careful con-
sultation with his advisors, including the joint Chiefs of Staff,
the President said he believed there was a way.

'Let me share with you a vision of the future which offers
hope. It is that we embark on a program to counter the
awesome Soviet missile threat with measures that are
defensive.

Let us turn to the very strength in technology that
spawned our great industrial bid and that has given us
the quality of life we enjoy today.

Up until now we have increasingly based our strategy
of deterrence upon the threat of retaliation.

But what if free people could live secure in the knowledge that their security did not rest upon the threat of
instant US retaliation to deter a Soviet attack: that we
could intercept and destroy strategic missiles before they
reached our own soil or that of our allies?

I know this is a formidable technical task, one that
may not be accomplished before the end of this century.
Yet current technology has attained a level of sophistication where it is reasonable for us to begin this effort.

And as we proceed we must remain constant in pre-
serving the nuclear deterrent and maintaining a solid
capability for flexible response.

But is it not worth every investment necessary to free
the world from the threat of nuclear war? We know it is!'[6]

The untold story of 'killer beams'

President Reagan didn't say it but the
race to perfect beam weapons was already in
full flight. In 1980, the United States spent
approximately $1000 million on beam
weapons research. The Soviet expenditure was estimated to
be three to five times that amount.

First to reach maturity are expected to be high-energy
lasers that concentrate energy with an extremely powerful
beam which, in the form of heat, can destroy a target
hundreds or even thousands of miles away. The beam strikes
almost instantaneously at more than 186 000 miles per second
but energy is lost when used in the atmosphere which causes
the beam to 'bloom' or defocus.

Another type is the charged particle beam — a stream of
subatomic particles of electrons, protons and heavier ions
accelerated to high velocities in an electromagnetic field.
Although less affected by weather, energy is lost by interaction with the air. There are also problems in aiming the
beam; the Earth's magnetic field causes it to bend and whip.
Scientists are now trying to produce 'neutralized' particle
beams which do not suffer this deficiency.

The story of the new superpowers' weapons race is
unfolding like a fictional space adventure. Information from

US observation satellites reveals the scope of Soviet ambitions:

■ At the Baikonur cosmodrome in central Asia the Soviets are preparing to test a powerful new space booster bigger than the Saturn V rocket which put US astronauts on the Moon. The US Defense Department claims that this rocket will lift into orbit six to seven times as much as the NASA Space Shuttle. It will launch a permanent space station which 'may support both defensive and offensive weapons in space with men in the space station for target selection, repairs, and adjustments and positive command and control.'

■ Research is underway on civil and military lasers at Troitsk, formerly Krasnaya Pahkra. Washington reports that lasers from this center may have damaged advance warning and reconnaissance satellites are unconfirmed. Almost certainly, they have been tested against the USSR's own Cosmos satellites.

■ A major research establishment south of Semipalatinsk is concerned with charged particle beams. This site was first described by Major General George Keegan, the former head of USAF Intelligence in 1977. Keegan claimed that the Soviets were embarked on a major research effort to develop the means of melting missile warheads by directed-energy beams.

■ An anti-missile test range has been constructed at Saryshagan, Kazakhstan, where prototype beam weapon equipment was installed in the late 1970s. Anticipated were the first Soviet tests of beam weapons against ballistic missiles launched from Kapustin Yar, east of Volgograd (formerly Stalingrad).

This is how the US defense establishment believes the Soviet ray weapons program will develop:

Mid-1980s. Moderate-power lasers for short-range ground-based applications such as tactical air defense and anti-personnel weapons.

Late 1980s — early 1990s. First prototype of a space-based laser.

Early 1990s. Operational system capable of attacking other satellites within a few thousand miles range.

1990s. Space-based testing of anti-ballistic missile (ABM) laser.

Early 21st century. Space-based ABM system operational.

Soviet goals and triumphs

As long ago as 1969, Leonid I Brezhnev gave an outline of Soviet future plans which parallels very closely events now unfolding. He said his country:

'was about to produce long-lived orbital stations and laboratories — the decisive means for the extensive con-

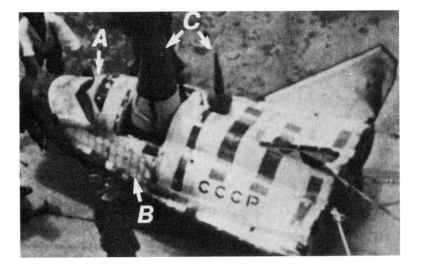

Above and right: The USSR's 'Kosmolyot' in miniature. A sub-scale prototype for a future manned space plane, it was recovered from the Indian Ocean by personnel from Soviet ships after making 1¼ Earth orbits. Revealing photographs, obtained by a P-3 surveillance aircraft of the Royal Australian Air Force in March 1983, show a vehicle resembling the USAF's X-24A lifting body (opposite page). Clearly seen is a small cabin. Thermal tiles cover the body for protection during re-entry. The cone-shaped object is an inflatable recovery aid. Flight testing of the 'Kosmolyot' research vehicle began in the mid-to-late 1970s with drop tests from a Tu-95 Bear over a Soviet test range. Three double Cosmos payloads launched by Proton D rockets in 1976, 1978 and 1979 were probably orbital re-entry tests for this program. In each case test vehicles returned to Soviet territory the same day. Then in 1982–83 came the orbital tests of the pilotless model featured here and launched by a two-stage booster from Kapustin Yar.

quest of space. Soviet science regards the establishment of orbital stations with replaceable crews as man's highway into space. They will become "cosmodromes" in orbit, launching pads for flights to other planets. There will appear large research laboratories for the study of space technology and biology, medicine and geophysics, astronomy and astrophysics.'[7]

Academician Boris Petrov, a former director of the Intercosmos program, spoke of new avenues of development which

The USAF's Martin Marietta X-24A lifting body. It first flew in 1969, anticipating the Space Shuttle by 12 years. The similarity of 'Kosmolyot' to the X-24A graphically illustrates just how far behind the Soviets are in this particular part of the space race.

space stations would open up in astronomy, biology, medicine and rocketry. Unique materials and precision parts, he said, could be manufactured under conditions of microgravity — and orbital stations would become launching pads for flights to other planets.

'At these stations final adjustments will be made to space-ships' systems; there will be training sessions and cosmonauts will be able to acclimatize to space conditions and participate in the assembly and checkout of interplanetary ships. From the station, too, laser as well as radio techniques can be developed for deep-space communications.'[8]

Academician Georgi Petrov, the former director of the Space Research Institute, mentioned the possibility of semi-automatic space stations in lunar orbit 'which could accept human visitors on fairly brief maintenance and research visits.'[9] Eventually, the Earth-orbiting and lunar stations could become terminals for ferrying materials for the construction of a lunar base.

We are now seeing the first fruits of this extensive program, which includes:

■ Multiple docking by manned Soyuz spacecraft with 19-tonne space stations Salyut 6 and Salyut 7.

■ Refuelling of Salyut 6 and Salyut 7 by Progress automatic cargo craft.

■ Testing of vehicles in orbit with multiple re-start, high-maneuver capability (eg, Cosmos 929, Cosmos 1001).

■ Engineering tests of 15-tonne modular units which can be docked automatically with a space station (Cosmos 1267 with Salyut 6, Cosmos 1443 with Salyut 7).

■ Preparation of launch complexes for super-boosters at the Baikonur cosmodrome.

■ Aerodynamic testing of a 'Kosmolyot' (spaceplane).

The 'Kosmolyot' will ferry cosmonauts between Earth and orbiting space stations. Although initially to be launched by a conventional rocket, it is planned to have an aircraft-type booster release it in the stratosphere. The mother will fly back to base to land on a runway.

The method allows a small spaceplane to be launched at any desired angle to the equator so that it can be over any spot on the Earth's surface within 100 minutes. The 'Kosmolyot' therefore could have great utility for making spot checks on military activities around the world before returning quickly to base. It might even be used in its own right for inspecting or disabling satellites in low orbits.

Satellite reconnaissance has now turned up evidence of a larger re-usable vehicle than 'Kosmolyot', resembling the NASA Space Shuttle but with a different engine arrangement.

In April 1983 this spaceplane was seen mounted on the back of a modified M-4 Bison bomber at the big experimental airfield at Ramenskoye which forms part of the Central Institute of Aero-Hydrodynamics (TsAGI) with extensive aeronautical laboratories and wind tunnels.

It has also been confirmed that new launch facilities 'are being built' at the Baikonur cosmodrome for the latest generation of Soviet heavy-lift space rockets.

Orbital battle stations and lasers

Pentagon officials were concerned that this expanding Soviet space program could lead to orbital battle stations moving out from a space station to intercept Western satellites. They could then attack with heat-sensing homing missiles or lasers.

The United States itself had not been idle. As long ago as 1973 two pilotless aircraft were shot down by an Air Force-developed gas-dynamic laser at the Sandia Optical Range, Kirtland Air Force Base, New Mexico. In one case the beam burned through control cables in the fuselage and on the other occasion it ignited fuel in the tank.

Three years later the US Army experimented with an Avco electric discharge laser fitted to a Marine Corps LVTP-7 amphibious assault vehicle. Winged and helicopter drones were destroyed.

Tests made in 1978 at a site near Camp Pendleton, San Juan, Capistrano, California, matched a chemical laser of moderate power against TOW anti-tank missiles. Each one, only 4 ft (1.2 m) long and moving at 450 mph (720 km/h), was destroyed.

Although these tests were all at short range, they demonstrated the big advantage of laser weapons over guided missiles: the ability of a laser to achieve fantastic rates of fire, perhaps a thousand times faster than any other weapon, with the ability to flick from target to target as the beam is shifted by a computer-controlled mirror-aiming device. However, the experimental unit, built jointly by TRW and Hughes, was far short of being a weapon. It was the size of a small factory.

A more powerful version of this laser is now being built in New Mexico with which experiments will be made for the US Air Force, Army and Navy. This 5-megawatt Alpha laser will be the most powerful to be demonstrated in the United States and could, in the future, be developed to produce more than twice that power.

In the meantime, tests have been made from the NKC-135 Airborne Laser Laboratory, a modified Boeing 707 airliner with laser fuel and equipment filling most of its fuselage. In June 1981, high above China Lake, California, the NKC-135 attempted to shoot down two inert Sidewinder air-to-air missiles passing at distances of 3-5 miles (5-8 km). They failed because of 'jitter' — the inability of the beam to remain focused on one spot on the target long enough to burn through.

The chief Western concern at present is that Soviet ray weapons soon may be able to disable satellites on which NATO depends for communications, reconnaissance, navigation, and ICBM early warning by directing bursts of laser energy at them. They might burn out sensors and complex electronic circuits. If such services were knocked out at the beginning of a nuclear attack, the whole allied defense network would be in disarray.

Consequently, American and British scientists are now working on methods of screening satellites against lasers. Protective measures include mirror-like surfaces from which beams bounce off, covers and 'glare' shields to protect sensitive equipment, and textured surfaces which diffuse laser light. Some satellites may have ablative coatings which melt when hit by a laser, dissipating the beam's energy before it can harm the main structure. Yet others may take the form of 'dark' satellites employing 'Stealth technology'; their casing materials absorb radar waves, making them poor tracking targets.

The danger is that the USSR — in the light of President Reagan's statement — will be spurred to step up its own considerable research on directed-radiation weapons, starting a new and enormously expensive spiral to the arms race. There is, after all, little chance of either superpower exchanging nuclear-armed ballistic missiles for a defensive screen of ray weapons this side of the 21st century.

X-ray lasers versus ICBMs Throughout history there has never been a defense system which approaches 100 per cent effectiveness, nor one that is invulnerable to countermeasures. Orbital beam weapons are unlikely to be the exception. Indeed, some of their features — as explained in this book — make them highly vulnerable to counterattack, not least by ray weapons themselves.

One US proposal involves a ring of 35 orbiting battle stations constantly alert for a missile attack. To be effective they would have to hit oncoming ballistic missiles within about eight minutes of launching, before they released multiple warheads and decoys.

Imagine a 21st century battle station with multiple lasing rods which discharge lethal pulses of penetrating X-rays, the energy coming from a small nuclear device. In theory several missiles can be engaged at the same time using UV-IR telescopes to track them in conjunction with an on-board computer to aim and fire. But as yet no one knows if a weapon of this kind can be made to work without destroying itself in the process. However, a crude X-ray laser device is reported to have been tested by the Lawrence Livermore National Laboratory at an underground test site in Nevada.

Enormous problems remain to be solved before such weapons can become a reality. Apart from getting the system to work with the necessary reliability, it must have an infallible advance warning system capable of identifying salvoes of missiles rising to the attack.

The technical problems in achieving this are immense. There is also the question of servicing the battle stations which presumably would have to be done by Space Shuttle astronauts. The cost would be exorbitant.

Another futuristic scheme envisages the launching of self-erecting 100 ft (30 m) mirrors which re-direct laser energy beamed at them from powerful projectors installed in high mountains. Launched 'when a nuclear war seems imminent' (to minimize the risk of an enemy knocking them out in advance) their prime objective would be to destroy Soviet missiles soon after launching.

Perhaps the answer may be found eventually in a two-tier defense system in which orbiting battle stations strike at ICBMs before they have time to release warheads, while a back-up system of ground-based charged particle beams catches missiles which elude them.

If this is the prospect that now confronts us, will it end the nuclear madness or merely extend the arms race? The kind of weapon described by President Reagan could serve the aggressor as much as the defender, simply because it reduces the power of a retaliatory strike. One has only to reverse the image and put the ray weapon shield into the hands of the USSR instead of the US. Unless both achieve equality in the new technologies, a serious destabilizing effect on the strategic balance may result.

The hope must be that possession of ray weapons — by both sides — will ultimately slow the development of ballistic missiles and steadily reduce the numbers deployed.

One thing is certain. It will be a long time before the threat of nuclear annihilation is lifted from the world by technology alone.

Chapter 1

MILITARY SPACE SYSTEMS

'The War of a community — of whole Nations, and particularly of civilized Nations — always starts from a political condition, and is called forth by a political motive. It is, therefore, a political act.'

Carl von Clausewitz,
On War (1832)

The battle for space was begun in earnest after World War II and is today a major priority of the superpowers. This extension of the rivalry between the US and the USSR into the zones above the Earth has political and Earthbound objectives. It commits the protagonists to diverse and expensive space programs, with military launches representing approximately one third of the total annual launches of the US and in the Soviet Union well over three quarters of the total annual figure.[1]

By means of their respective satellite networks, each superpower monitors the forces and crosschecks the claims of its rival with rigorous accuracy. Bluff and counter-bluff belong to the past. Each knows the nature, number and location of the other's forces and the paradoxical result is that spy satellites tend to have a stabilizing effect on world affairs.

Photo-reconnaissance satellites have gone through several generations to reach their present level of sophistication. The US currently uses two different types. One is the Big Bird, first launched in June 1971.

It lives up to its name, being approximately 50 ft (15.24 m) long and 10 ft (3.05 m) in diameter. It weighs roughly 30 000 lb (13 608 kg), and is launched by Titan III-D from Vandenberg Air Force Base into a 155 by 100-mile (250 by 160-km) Sun-synchronous orbit.

The Big Bird represents a milestone in that it combines functions previously requiring two different types of satellites — search-and-find (locating new targets through a wide area surveillance) and close-look (the examination of these targets with high-resolution optics). Film from the search-and-find cameras is developed on board, scanned and then transmitted via radio to a ground station. If something of interest is discovered, it can be re-photographed by the high-resolution system within a few days. This huge camera, built by Perkin Elmers, makes up most of the payload and has a resolution of better than 1 ft (0.3 m) so that it is capable (for instance) of identifying personnel and vehicles on an airstrip.

It is necessary to return the film to Earth to achieve these results since the critical, fine details this camera can provide would be lost if the alternative method of radio digital transmission were used instead. Four re-entry capsules are available and, at predetermined intervals, one is loaded with exposed film and de-orbited, to be caught in mid-air by specially-equipped Air Force HC-130 aircraft of the 6593 Test Squadron flying out of Hawaii. A trapeze-like rig trails behind the aircraft and this rig snags the parachute; the capsule is then reeled in like a fish on a line. The spacecraft is believed to have a multi-spectral capacity for detecting camouflaged and underground installations like missile silos and buried command centers, and an infrared scanner for monitoring night activity. Approaching the end of its long, productive life, the Big Bird series is due to be phased out in 1984.

Looking through the Key Hole

The other satellite is the revolutionary Key Hole-11, first launched in December 1976. The Key Hole uses the same launch vehicle and launch site as the Big Bird but is put into a higher orbit — 310 by 155 miles (500 by 250 km). It differs from all previous reconnaissance satellites in that it dispenses with film in favor of digital signal transmission. The camera system is designed around an array of light-sensitive silicon diodes on a principle similar to that of a photographer's light meter. Electronic signals are read off and transmitted to Earth. Depending on the strength of the signal from each element in the array, one of several hundred shades between pure white and jet black is assigned. The picture itself is made up of a series of these tones. The advantage of this technique is that Key Hole-11 is not limited by the amount of film the satellite can carry. The Big Bird could stay aloft for about six months; the first Key Hole-11 lasted 25 months. What is more, the photos are returned on a near-live basis whereas previously a delay of several weeks might intervene. The system's only shortcoming is that it lacks the excellent resolution of the Big Bird.

The importance of these space systems is brought home when it is realized that several advanced reconnaissance satellites are in secret storage reserved for use in a national emergency. Their resolution from 70 miles (112 km) altitude is perhaps 6 in (15.2 cm), probably enough to discriminate between different makes of cars on a Moscow street.

Under development for launch in the 1980s are several even more sophisticated reconnaissance satellites. A modified

A2 SOYUZ

Key Hole-11 is planned for deployment in 1984 which has a resolution similar to that of Big Bird despite the use of digital transmission. It can be launched by either a Titan 34D or the Space Shuttle. Later models will be recovered by the Space Shuttle for refurbishment after they have exhausted their fuel or other consumables.

Discussions have taken place about equipping the Space Shuttle's payload bay with high-resolution cameras for photographing high-priority targets. This is another proposal that would use film-type cameras and as before re-entry capsules would be used to return the film. It could be done on only a limited basis and the scheme is still somewhat tentative.[2]

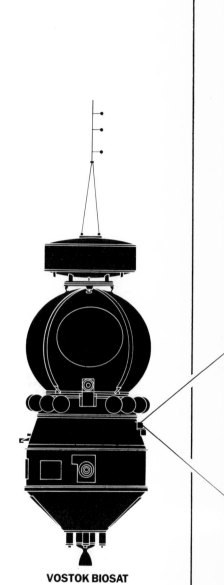

VOSTOK BIOSAT

Eyes of the Soviet Union

The Soviets have a fundamentally different operational procedure for their reconnaissance satellites. In sharp contrast to the US emphasis on building a few satellites with very long lifetimes, the Soviets launch large numbers of satellites with a more limited time in orbit. Although many more Soviet satellites are launched, their actual effectiveness is probably on a par with the US. Often two or three Soviet reconnaissance satellites are in orbit simultaneously, all flown under the blanket Cosmos label.

From the Cosmos program's inception in 1962 right up to the present day, the Soviets have used modified Vostock spacecraft. The satellites weigh around 14 000 lb (6350 kg) — about the same as a medium-sized truck — and are launched by A-2 boosters. The Vostok was the pioneering Soviet spacecraft which put Gagarin into orbit, and the series was originally designed for automatic operation with the cosmonaut as a passenger. In the reconnaissance version, camera and film apparently replace the cosmonaut and ejector seat inside the spherical re-entry capsule. That 20 years has seen almost 600 reconnaissance satellite launches is a remarkable fact.

Soviet reconnaissance satellites are divided into two types. First is the maneuverable high-resolution satellite. On top of the re-entry sphere is a rocket engine used to adjust the ground track. The satellite makes repeated passes over the target area at the same time each day, thereby achieving identical lighting conditions and making interpretation of the photos easier. This capability was used to advantage during the 1973 Middle East War and subsequent crises. A typical orbit is 154 by 130 miles (248 by 210 km), although on occasions this can vary widely. Typically a satellite stays in orbit 14 days.

The other satellite type, equipped with lower resolution optics, is used for search-and-find activities. Like the close-look, this version spends 14 days in orbit before retro-fire and landing in the Kazakhstan recovery zone. Satellites of this

The Battlefield of Space

When the USSR put Sputnik, the first man-made satellite, into orbit on October 4, 1957, humanity crossed an important threshold. Today, the zones around the Earth are 'mapped out' for both civilian and military space activities. Specific orbits are chosen for specific tasks — from the reconnaissance satellites whose camera-based espionage calls for relatively low altitude, to invaluable communications satellites spaced around the Earth's equator in high geostationary orbits.

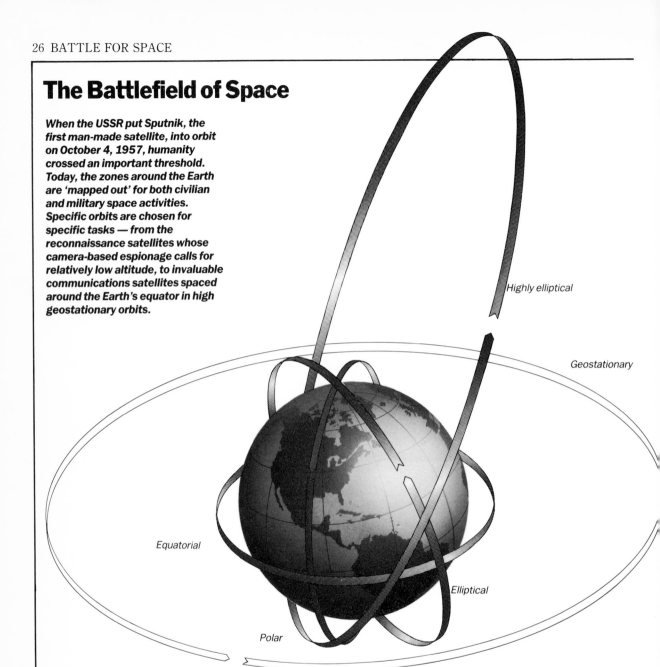

Highly elliptical

Geostationary

Equatorial

Elliptical

Polar

Above: An orbit is the path in which a body moves relative to a source of gravity. The plane of any satellite orbit bisects the Earth through its center.

A satellite travelling around the Earth at a constant speed and on a path at all times the same distance from the Earth's center of gravity follows a circular orbit. A circular orbit will become an ellipse with the slightest change in velocity or direction.

A satellite is in an elliptical orbit when travelling a closed path which is longer than it is wide. The key points in the satellite ellipse are apogee, the point farthest from the Earth, and perigee, the point closest to the Earth. The advantages of this kind of orbit are illustrated by Soviet early warning satellites which use a highly-inclined 12-hour elliptical orbit with its high point over the northern hemisphere. This allows infrared sensors aboard the satellite to take in the whole of the North American continent, with a five or six hour observational period over the US every day.

Geosynchronous orbit has the advantage that the velocity of a satellite in it matches the Earth's rotation. It seems to 'hover' in the sky as seen from the ground. This orbit was used first for communications satellites after a suggestion by Arthur C Clarke, then later for early warning and ELINT satellites.

A satellite can be launched in many directions from a given point on Earth. The choice of angle or inclination depends on the intended use of the satellite. Different inclinations allow a satellite to cover different areas of the Earth's surface. A satellite fired directly north or south could, in a given period of time, look down on all parts of the globe, since the Earth turns independently inside the ring described by the satellite.

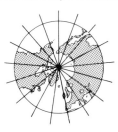

A satellite fired due east at 30° north latitude would not follow the 30th parallel around the globe, but would adjust itself to a plane cutting through the Earth's center of gravity.

Right: The battlefield of space has two important regions, given current technology. The first and most important is low Earth orbit — from the top of the atmosphere up to perhaps 700 miles (1126 km) altitude. This at present encompasses the orbits of virtually all military satellites — from low-altitude reconnaissance satellites up to weather and navigation satellites.

The other vital region is the band of geosynchronous orbit, 22 300 miles (35 880 km) high, the home of early warning satellites and electronic 'ferrets'. As more and more missions are transferred, geosynchronous orbit could in the future become as important as low orbit. The region between these two bands of activity is not used very much at present. It is, however, the orbital territory in which future laser battle stations may operate.

22 300 miles
Geostationary
orbit

Inclination and ground tracking

If a satellite's orbit is to cover most of the Earth's surface, it must be fired at a fairly sharp angle to the plane of the Earth's equator. For example, a 45° shot (at either a northern or southern angle) describes a crisscross path over the Earth's rotating surface between 45° north latitude and 45° south latitude. This orbit can be created only from a launch base between these latitudes.

A 65° angle, Sputnik's inclination (see diagram), extends coverage to

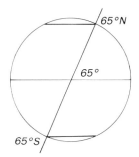

mid-Alaska in the north and almost to the Antarctic continent in the south, as this ground track shows.

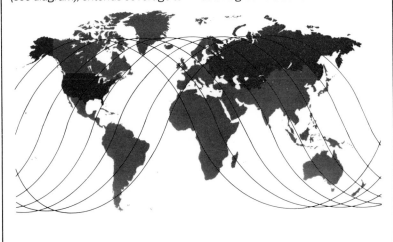

700 miles

110 miles
Upper reaches of
the atmosphere

The enormous distances spanned by the battlefield of space are represented to scale here with the Earth's atmosphere compressed into less than one millimeter.

Ground level

second type are placed into a 258 by 223-mile (415 by 360-km) orbit, and the large film capsule is not caught in mid-air but recovered by a special team equipped with helicopters and ground vehicles.

The Soviet reconnaissance satellite program is a dynamic effort with many offshoots and frequent technical improvements. During the mid-1970s for example, the low-resolution satellites lacked a maneuvering engine. Now most Soviet reconnaissance satellites have them. If the satellite malfunctions in orbit, it carries a self-destruct package to prevent uncontrolled re-entry, so precluding the possibility of capsule and camera falling into unfriendly hands. Several have in fact been exploded in orbit.[3] Some Soviet reconnaissance satellites are believed to carry Earth-resources packages, performing such tasks as scanning the oceans for plankton and measuring wave heights. Their orbits are somewhat higher than those of the high-resolution reconnaissance satellites; typically 170 by 164 miles (275 by 265 km).

The USSR has been slow to field a replacement for the Vostok-based reconnaissance satellites. The new type is believed to be based on the Soyuz manned spacecraft. The vehicle in question is conjectured to have two re-entry capsules instead of the orbital module which in a manned mission carries supplies and equipment. These are de-orbited at different times during the mission to provide information. When the mission is completed, the main capsule containing the camera is recovered. The first such satellite was Cosmos 758 in September 1975 which evidently suffered a malfunction and was destroyed. Originally, the Soyuz-based reconnaissance satellites stayed aloft for 29.5 days, but in recent years this has been extended to as much as 2 months. The long lifetime of this satellite is perhaps due to the use of solar panels rather than the chemical batteries of the Vostok reconnaissance versions.[4]

Launches were very sporadic in the early years of the program, but the number has increased to 7 or 8 per year in the 1980s. They are placed into 218 by 106-mile (352 by 171-km) orbits, the low altitude indicating that they are high-resolution satellites (since resolution diminishes as height from Earth is increased).

The Soviet Union experimented with manned photo-reconnaissance missions in the mid-1970s under the guise of Salyut. The enigmatic Salyut space station effort actually comprises two separate programs, one civilian, the other military, though these distinctions are rarely clear-cut. The first 'military' station was the ill-fated Salyut 2 launched in April 1973. In conjunction with the military-civilian station effort, it was regarded by Western military observers as a bid to upstage the US Skylab program by having two space stations in orbit simultaneously. If so, the bid failed as both suffered disastrous malfunctions. Salyut 2 started to tumble and broke up. The civilian station was disabled at launch. To hide the failure, it was given the cover name Cosmos 557.

Soviet military space stations Subsequent Salyut flights — 3,4,5 and 6 — had a very different character.[5] Salyuts 3 and 5 were military stations. Unlike their civilian counterparts, they employed telemetry frequencies and formats previously associated with reconnaissance satellites. The spacecraft commander and flight engineer were both military officers and the stations were put into approximately 150-mile orbits as against the 217-mile (241 against 350-km) orbit of the civilian stations. Such a low orbit requires more frequent corrections to prevent decay but provides better ground resolution.[6,7] The stations bore a 33-ft (10-m) focal length camera; ostensibly a solar telescope, although the optical system is better suited to reconnaissance. Ground resolution might be as good as 1 ft (0.3 m), depending on the film.[8] Film and magnetic tape are returned to earth by a recovery capsule after the station has completed its mission.

These military stations differ in design from the civilian stations in a number of ways. The Soyuz transport spacecraft docks at the aft end of the station (like Salyut 6). The solar cell panels are farther aft and there are only two pressurized sections. The Soviets have been reluctant to divulge information about these missions; almost no television views of the interior exist and Soviet media coverage was spotty.

The current status of the military Salyut program is uncertain. A total of five manned missions were launched toward the two stations and the flights achieved mixed results. Two failed to dock and were aborted. Another manned mission was reportedly cut short. The program was part of the renewed Soviet space activity in the wake of Apollo and the deaths of the crew on the Salyut 1-Soyuz 11 mission.

At present, the Soviets are striving to establish a permanent space station. Given the major effort they have made over the years in military space flight, it seems likely that military goals will figure in its design. The technology demonstrated during the civilian Salyut 6 mission would serve any future military station. The Progress resupply spacecraft could shuttle film and equipment, and the manifest ability to refuel a station means that a low orbit would not be as difficult to maintain. A reuseable shuttle, which the Soviets are working on, could put the rotation of crews on a more cost-effective basis.

The People's Republic of China also has a space reconnaissance program which began in 1975. The satellites weigh about 6000 to 10 000 lb (2721 to 4536 kg). After completion of each mission, a large re-entry capsule separates, fires a retro-rocket and returns to Earth. A module is left in orbit. The reconnaissance satellite is an important military capability for the Chinese because of the Soviet build up on their border. Unlike the US and Soviet efforts, the Chinese space program is highly sporadic and on occasion years have passed between launches.

Electronic 'ferrets'

Photo-reconnaissance is not the only type of surveillance activity; there is ELINT (electronic intelligence) as well. ELINT refers to the reception of electronic signals such as radar, radio and telemetry from enemy installations which are located by receivers aboard the relevant satellites and these satellites are popularly known as electronic 'ferrets'. From such recordings, the characteristics and area covered by a given installation can be determined. The important characteristics include transmission frequency, pulse duration, antenna beam pattern and polarization, so that ELINT essentially means listening rather than looking from space.

The information they provide is used in the design of jamming equipment and decoys for bombers and missile warheads. By knowing the precise characteristics of an enemy's monitoring systems, it is possible to evade detection by flying through gaps and weak spots in his coverage. Radio transmissions can indicate the location and identity of military units, their equipment, readiness, morale and operating procedures.[9] Analysis of radio traffic gave the first indication that the USSR was developing ICBMs in the early 1950s; and analysis of telemetry is now routinely used to determine the configuration and performance of Soviet missiles.

At present the US employs an ELINT system code named Rhyolite. The satellite enters a 22 300-mile (35 880-km) high circular orbit above the equator. It takes 24 hours to complete this orbit and appears to hover over one spot on earth. Rhyolite carries a virtual forest of antennas, including a dish antenna at least 70 ft (21.3 m) in diameter.[10] Its onboard equipment can isolate the desired transmission from the hash of earthly background noise. The first Rhyolite was launched in March 1973 and by 1983 two were stationed above the horn of Africa to record telemetry from liquid-fueled ICBMs launched on test from Tyuratam towards the Kamchatka Peninsula. Another pair is positioned to monitor the launch of missiles from Plesetsk — the military launch centre in the northwest USSR.

Small ELINT packages are also launched 'piggyback' aboard US Big Bird reconnaissance satellites. Once the Big Bird is in orbit, the ELINT package separates and fires its onboard engine to propel it into a higher orbit. Those placed in a 300-mile (482-km) orbit are for monitoring radar transmissions. A few are placed in 900-mile (1448-km) high orbits; evidently to monitor anti-ballistic missile radar units. These ELINT cargo sub-satellites are regular features of Big Bird

Different types of reconnaissance. The spyplane (left) is highly mobile but at risk from enemy forces if detected. Certain reconnaissance satellites (center), operating at a relatively low altitude, may obtain high-resolution pictures on each sweep over the target location, even through cloudcover, but cannot supply uninterrupted coverage. A satellite in geosynchronous orbit (right) may cover the target location continuously but tends to lack the excellent detail attainable from a lower altitude. Recent technological improvements have narrowed the gap, however.

launches and their purpose is to conduct a general survey; that is, to detect any new radar sites or changes at established ones.

A successor to Rhyolite, codenamed Argus after the monster of Greek legend, was proposed only to be vetoed by Congress in 1975 in a dispute over its value. Ironically, if Argus had gone ahead, it would have been ready in time to replace the ground stations closed by the Iranian revolution. The Tacksman 2 facility in northeast Iran formerly monitored launch activities at Tyuratam, tracking the telemetry of Soviet missiles from launch through mid-trajectory. The vulnerability of ground stations to political upheavals has made a Rhyolite successor an even higher priority. A satellite codenamed Aquacade is now under development, designed to allow launch by Titan 34D or to be actually placed in space by the Space Shuttle.

Drilling for radioactive debris from the nuclear-powered Cosmos 954 in Canada.

The availability of the Shuttle will allow the orbital assembly of extremely large antennas for ELINT and early warning functions. Diameters of several hundred feet are thought possible. Such large antennas are necessary as the great distance between the Earth and the geosynchronous orbit they would occupy weakens the already faint signal from a ground source or an ascending ICBM. Ground stations on the other hand only have to cope with distances of a few hundred miles.

The Soviets use two different types of ELINT satellites. The first, launched about 4 times a year by a C-1 booster into a 340-mile (550-km) orbit, very likely conducts search-and-find activities. They form a regular network in orbit and launches are timed so that the functioning satellites are spaced at 45° intervals around the globe.[11] A larger ELINT package is launched about once a year into a 370-mile (600-km) circular orbit by an A-1 booster. With a payload three or four times that of the C-1 variety, these satellites can conduct a close examination of a newly detected radar site. The first of this type was Cosmos 389, launched in December 1970.[12]

Ocean surveillance: nuclear and ELINT

Since the 1960s, the Soviet Navy has grown phenomenally. Formerly little more than a coastal defense force, its ships now challenge US naval dominance in all the world's oceans. In support of these fleet activities the Soviets fly two types of ocean surveillance satellites. The first is a nuclear-powered radar-equipped satellite — launched by an F-1-m — which employs an unusual flight profile. Two satellites are launched within a few days of each other into a 168 by 159-mile (270 by 256-km) circular orbit. The satellite is about 46 ft (14 m) long and 6.5 ft (2 m) in diameter. Equipped with its own radar, it can detect shipping regardless of weather conditions or the ships maintaining radio silence. The radar unit is powered by a nuclear reactor carrying about 110 lb (50 kg) of enriched uranium.

Twin satellites work together to track Western shipping and US-carrier task forces in particular. The satellites are kept one complete revolution apart or with one satellite following a few minutes behind the other. Small orbital adjustments maintain this spacing. After 60 to 75 days, the reactor unit separates and is boosted into a 590-mile (950-km) orbit, not to re-enter the earth's atmosphere for 600 or more years and hopefully after the radio-active material has become harmless.

The first such flight was Cosmos 198 in December 1967. Development was slow and the first dual flight did not come until May 1974 with Cosmos 651 and 654.[13] This type of satellite was forcefully brought to public attention when Cosmos 954 suffered a malfunction that prevented the reactor from

being fired into the higher orbit. Atmospheric drag slowly brought it lower, until in January 1978 it re-entered, scattering radioactive debris across the Canadian Northwest Territories. A multi-million-dollar clean-up effort was required for which the USSR subsequently reimbursed Canada. Cosmos 1176 was launched, after an interval of two years, in April 1980 and the program goes on.

The other and less controversial Soviet satellite employs ELINT techniques. The first was Cosmos 699 in December 1974.[14] Radio and radar transmissions from sea-going vessels are used by this satellite to plot a ship's location and type. Signals from an aircraft carrier will differ from those of a destroyer since their radar systems and modes of operation are different. Even individual ships in the same class have idiosyncrasies that allow identification. The pattern was that a single mission would be launched in the interval between pairs of the nuclear-powered satellites. But during the hiatus in nuclear satellite launches, a pair of ELINT satellites went up — Cosmos 1094 and 1096. ELINT employs a 280 by 267-mile (470 by 430-km) orbit, which is higher than the nuclear type.[15]

Published for the first time in the West, a Soviet artist's impression of Vostok 1. In the reconnaissance version camera and film replace the cosmonaut and ejector seat. Such reconnaissance satellites are still in use, alongside a new type based on the Soyuz manned spacecraft. The latter has a longer lifetime, possibly because it is equipped with solar panels.

Secret Life of the Salyut

This picture sequence of a civilian Salyut space station offers clues to the controversial Salyut military missions. A 33 ft (10.05 m) focal length camera replaces the telescope shown in the cutaway. Two cosmonauts operate the reconnaissance camera and give a verbal report on what they see. Film is returned at the mission's end by a separate recovery capsule.

To test crew and camera, 'objects' were set out near the Tyuratam launch site: eg, resolution-testing charts, aircraft and tanks. (Ships, aircraft in flight and buildings can be seen from orbit with the naked eye.)

Salyut 5 evidently took part in military exercises in July 1976, monitoring the Eastern Siberia – Northern Pacific areas. Its crew supplied real time tactical information to battlefield commanders. Three military Salyuts have gone up — Salyut 2 (which failed after launch), Salyut 3 and Salyut 5.

1 Propulsion system
2 Progress service module
3 Progress – Salyut docking system
4 Docking hatch
5 Rendezvous antennae
6 Main propulsion engine
7 Telescope and cryogenic cooling unit
8 Running belt
9 Steerable solar array
10 Soyuz – Salyut docking system
11 Soyuz instrument module
12 Soyuz descent module

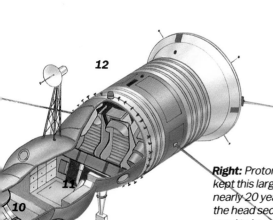

Right: Proton D. The Soviets have kept this large rocket under wraps for nearly 20 years. This picture shows the head section of the D. variant used to launch Soviet Salyut space stations. The rocket is in the act of launching the 18.5 tonne Salyut 1 at the Baikonur cosmodrome on April 19, 1971. The picture, taken from a television screen in the Soviet Union, was deliberately cut as the vehicle rose from the launch pad and began to reveal nosecaps of six strap-on boosters. Despite the secrecy, Western Intelligence has been able to examine the rocket from satellite photographs.

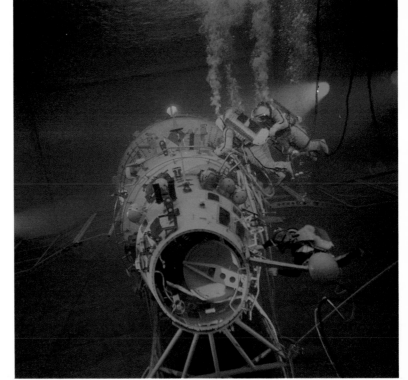

Right: Salyut cosmonauts practice spacewalking in a water tank at the Hydro-Laboratory of the Yuri Gagarin Cosmonauts' Training Center outside of Moscow. The tank gives cosmonauts a 'feel' for spacewalking on the outside of a space structure. In this way it has been possible to simulate repair and maintenance tasks under conditions as close to weightlessness as it is possible to achieve in a ground laboratory. Two cosmonauts, suitably weighted to ensure they neither rise nor fall in the water, are in the company of divers who look after their safety.

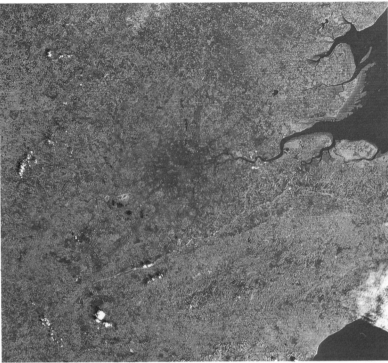

Above: A Titan IIID lifts off with Big Bird aboard. No photos have been released, but Big Bird reportedly lives up to its name – 50 ft (15.2 m) long and 10 ft (3 m) in diameter. It is said to have six film recovery capsules.
Above right: London and the Thames estuary as seen from Landsat, NASA's Earth resources satellite.

The US was slower to fly ocean surveillance satellites, relying instead on wide-ranging aircraft carriers, land-based patrol aircraft, radio direction finding and networks of underwater sound detectors. Ocean surveillance satellites were a high priority for the Soviets because of the USSR's few ice-free ports and limited access to the open oceans. The US has the geographical advantage and so such satellites were less important. The Navy's initial experience with satellite-based ocean surveillance came from satellites launched by the Air Force using Titan IIIB-Agenas in the early 1970s.[16] The operational system which ultimately emerged, codenamed White Cloud, employs a cluster of multiple satellites launched by a single Atlas F. These satellites are put into parallel orbits so that spacing is maintained, and a test of the cluster system was made in December 1971.

The first operational cluster was launched in April 1976 into a 1815 by 1757-mile (1128 by 1092-km) orbit. Probably equipped with multiple sensors, the satellites have ELINT equipment for listening to shipboard transmissions. The various bearings from elements of the cluster precisely locate the source.[17] An infrared sensor is evidently part of the package as well. A warm ship against a cold ocean background makes an excellent infrared target, and warm water discharged from a nuclear submarine's reactor can also be detected from space.[18]

Unlike photo-reconnaissance of land targets, ocean surveillance is predominantly tactical. Ships, after all, can move over vast distances in a day. A premium is therefore placed on the quick reception, analysis and dissemination of ship position reports if they are to be useful.

Future US ocean surveillance satellites will use active radar. One such satellite, codenamed Clipper Bow, failed to win approval for development in 1979 due to its high cost and what was deemed limited value. Its successor, now in the study phase, is the Integrated Tactical Surveillance System. The ITSS uses radar but has many more applications besides ocean surveillance. The Army and Marine Corps are involved and the Air Force also may use ITSS to detect enemy aircraft, demonstrating the US philosophy of using a small number of satellites for many purposes.[19]

Early warning satellites

The Soviet development of the SS-6 ICBM in the late 1950s presented the US with a difficult new defense problem. For reasons of political geography, warning time of a Soviet ICBM attack amounted to only 15 minutes, since allied radar stations in Greenland, Alaska and England could not detect a missile until it came over a station's horizon (some 15 minutes after launch). This meant that in theory US nuclear forces could be destroyed on the ground.[20] The argument ran that if warning time could be increased, more US bombers could get off the ground and, with a greater chance of a counterstrike, a surprise attack would be less likely. It was now recognized that a satellite carrying infrared sensors specifically designed to detect an ICBMs exhaust plume would provide 30 minutes' warning.

Despite the high priority of the mission, US early warning satellites had a sporadic development. The original program was called MIDAS (Missile Defense Alarm System). Cut back in mid-1962 to a research and development program due to poor reliability, it also suffered when the Soviet missile threat did not develop as quickly as was first feared. MIDAS was then followed in 1968 by Program 949, which tested features of the operational system such as use of inclined geosynchronous orbit and improved sensors. This was an important experimental area, as an inclined orbit allowed the satellite to trace out a figure-8 and so gave it better coverage of the USSR landmass. Both programs relied on the Atlas-Agena launch vehicle.[21]

Undergoing bureaucratic metamorphosis, the operational early warning system was originally known as Program 647, later as the Integrated Missile Early Warning Satellite and finally as the Defense Support Program. A guidance system malfunction left the first satellite stranded in a 22 300 by 16 190-mile (35 886 by 26 050-km) orbit, after its launch by a Titan III C in November 1970. Only limited sensor testing was possible. The first successful Program 647 launch came in

May 1971 when the satellite achieved a geosynchronous orbit. Flights have been made at roughly one year intervals since.

The operational system consists of three satellites. One is stationed over the Indian Ocean to monitor launches from the USSR and China. Two are in position over the Pacific and South America respectively to detect submarine-launched missiles. The Program 647 satellites carry a 11.9-ft (3.63-m) long infrared Schmidt telescope with an aperture of 3 ft (0.91 m). It is angled to the cylindrical equipment section by 7.5°. At the telescope's focus is an array of 2000 lead sulphide cells, each of which scans a region on earth less than 1.8 miles (3 km) across.[22]

A cell's sensitivity can be changed as an aid to identification of an infrared source — that is, to ascertain whether it is a rocket or a ground source (such as a forest fire) or the sun reflecting off high-altitude clouds. (The latter was one source of MIDAS' chronic false alarms).

The satellite spins at a rate of 5 to 7 revolutions per minute and its telescope scans a particular location on the earth every 8 to 12 seconds. Plotting an infrared source's position over a number of scans determines whether it is from a missile in flight or a stationary source. Identification of the source requires 50 to 60 seconds and warning of an attack can be relayed to Earth within 90 seconds of missile liftoff.[23,24] Signals are radioed to a control center near Alice Springs, Australia or to a similar facility at Buckley Field, Colorado, then to the NORAD and SAC Headquarters, and small ground stations are planned. Missiles can be tracked until final-stage burnout, although they would then be in range of early warning radars.

A refined version of the Program 647 satellite was first launched in June 1973. Nuclear-explosion detectors were included to monitor the test ban treaty, and provide valuable data on cosmic and X-rays outside the Earth's atmosphere.

Early warning satellites are of course flown by the Soviet Union as well and the same taxing technical problems had to be overcome. Soviet early warning satellites use a highly inclined 12-hour elliptical orbit — 24 240 by 373 miles (39 000 by 600 km) — with high point over the northern hemisphere. Infrared sensors aboard the satellite can take in the whole of the North American continent. Its orbit allows a 5 or 6-hour observational period over the US every day and the Earth's rotation causes the high point to occur over the USSR on the satellite's second orbit*.

Cosmos 159 in May 1967 was the first of the batch. However, between December 1968 and September 1972, no Soviet

---

*This unusual orbit is used because of difficulties in achieving a geosynchronous one from so northerly a launch site. Cosmos 775 was ambitiously placed in geosynchronous orbit during October 1975, perhaps to test a new strain of early warning satellite. Since then, however, it seems there have been no successors[25,26].

early warning satellites were launched; possibly because, like MIDAS, they suffered from false alarms and poor reliability. Not until 1977 did the satellites achieve even partially operational status.

Weather and navigation satellites

Reconnaissance, early warning and ocean surveillance are not the only military missions undertaken in space, and the two most important 'semi-civil' missions are possibly weather and navigation. The primary military function of weather satellites is to support reconnaissance satellite activity, since planners must know if the area to be photographed will be clear of clouds. Even shadows from patchy clouds can make analysis difficult, to say nothing of the wasted film, time and effort.[27]

During the early 1960s, the Air Force employed civilian weather satellites but civilian and military goals diverged and in January 1965 the Air Force began to fly a separate weather satellite program. Civilian weather forecasters wanted a broad geographical overview, whereas the Air Force wanted detailed information on specific areas.[28] Moreover, the civilian system could not meet the exacting Air Force accuracy requirements. (Military weather satellite photos incidentally were not made available for civilian use until March 1973).

The Block 5D weather satellite flown in the late 1970s is equipped to provide cloudcover photos in both visible light and infrared as well as temperature and water vapor measurements. Its photos have two resolutions; 2 and 0.33 nautical miles (3.7 and 0.61 km), nearly twice the available resolution from civil weather satellites at the time. Its basic design provided the blueprint for NOAA's answer to the dove which Noah released from the ark: the Tiros N weather satellite.

NOAA (the National Oceanic and Atmospheric Administration) drew on the technology which the Air Force had developed to provide higher-quality data and imagery. Launched by a Thor rocket into a 500-mile (805-km) Sun-synchronous orbit, the Block 5D satellites transmit signals to Air Force ground stations which are then relayed to the Air Force Global Weather Center where they are processed automatically by digital computers. Airmobile vans equipped for direct reception of photos at forward locations were used during the Vietnam war to provide weather data for bombing missions over the north.

The USSR was slower to fly weather satellites, and the program when it came was not divided into military and civilian parts. This followed the general Soviet principle of all space developments having a simultaneous military and civilian use. Two or three Meteor 2s are operational at a time and they are launched about four times each year. They are

A military satellite also available for civilian use, the US Air Force's Block 5D assists reconnaissance missions by supplying data on cloudcover. Reconnaissance satellites can then take advantage of the best conditions.

The nuclear-armed Polaris gave new urgency to navigation satellite research. The location of the submarine launching it had to be known precisely in order for it to hit the target.

spaced at intervals of 90° to 180° apart which enables each satellite to record weather conditions over a given spot on Earth every 6 to 12 hours.

Meteor 2 satellites are launched by an A-1 booster into a 557 by 521-mile (897 by 839-km) orbit, similar to US Air Force satellites. When the first Meteor 1 was launched (a prototype of the present satellite) on March 26, 1969, it was described as having a resolution three times that of NOAA's Tiros satellites, which is enough to support reconnaissance satellite activities.

The Meteor 2s, equipped with a cloudcover television system and an infrared radiometer for night cloud imagery, also carry systems for measuring moisture content and the vertical temperature profile of the atmosphere. They transmit data to over 50 stations in the USSR which the Soviet Hydrometeorological Service processes in about 90 minutes. The Meteor satellites have greatly improved weather forecasting, their data allowing considerable savings in such diverse fields as irrigation, shipping and air travel. Sporadic tests of weather satellite systems were conducted by the Soviets between 1963 and 1965 and regular satellite weather data was not established until 1966 or 1967.

The dawn of satellite navigation

Satellite navigation was first suggested in 1869 by Edward Everett Hale in *The Brick Moon*. He envisioned the use of four artificial satellites to determine longitude, something that had been impossible until the development of the chronometer in the early 18th century. Since then radio techniques such as direction finding to shore stations and Decca, Loran and Omega navigation systems have entered the picture. Nevertheless, just as celestial navigation can be defeated by clouds, so these radio systems can be disturbed by local interference and remoteness from the station. Except in coastal waters, accuracy diminishes with distance from the shore station.

Nothing better illustrates the political dimension of the battle for space than the advent of the missile-launching Polaris submarine. Polaris required the development of precise, all-weather techniques since if the missiles were to hit their targets the launch point had to be known with great precision; a difficult task when the launch platform was constantly in motion.

Navigation satellite research consequently became an important priority. The world's first navigation satellite was the Transit 1-B launched in April 1960 by a four-stage Scout into a 700-mile (1126-km) circular orbit and the series has since undergone steady improvements (the latest version is known as Nova).

Transit satellites carry two ultra-stable oscillators and

two dual-frequency transmitters which allow relatively accurate pinpointing. Once the satellite's orbit is precisely known, a location on Earth can be determined from the Doppler shift in transmitter frequency caused by the satellite's orbital motion.[29] For a single pass accuracy is to within about 200 ft (60 m) if a ship's speed is known exactly, but drops to 600 ft (183 m) if the velocity is in error by even 0.5 knots (0.93 km). On land, by comparison, multiple passes help to locate a fixed point within 15 ft (4.6 m).

Transit satellites are limited as well with regard to tracking faster-moving vehicles such as aircraft. After a slow start when the system was released for general use in 1967, the number of civilian users has increased so considerably that they now total four times the number of naval users.[30]

The Soviets, like the US, built missile submarines; first the Gulf and Hotel classes (as they are called by the Western military), then the Yankee, Delta and most recently the Typhoon. As with the US Polaris submarine, these require support from navigation satellites. The first such satellite was Cosmos 192, launched by a C-1 in November 1967.

In general, Soviet navigation satellites are launched 4 times per year into 622-mile (1000-km) high polar orbits. Two separate networks have been established which are slightly offset from each other. Frequencies and techniques are identical to those used by the US Navy's Transit and the popular name for the Soviet system is Cicada (since the orbital radio beacons sound like a chorus of insects).[31,32]

After a long period of study, test flights of the Timation satellites began, leading to the Navstar Global Positioning System. The first example was launched from Vandenberg by an Atlas F in February 1978. Navstar's orbit is approximately circular at 12 550 miles (20 200 km); it has a period of 12 hours. This system uses a method called time ranging to help a user find his bearings, the user's distance from the known location of the Navstar being determined from a measurement

A Soviet Delta-class ballistic missile submarine, dependent like its US counterparts on navigation satellite support.

Above: Artist's impression of a Navstar. The atomic clocks which each satellite carries are so accurate that Relativity must be taken into account when using them.
Below: The Navstar network of 18 satellites all in polar orbits. Signals from these satellites are used to pinpoint the user's position.

of the time taken by the signal to cover that distance. Signals from three satellites are needed to pinpoint a position. The time ranging system, simple as it is in principle, was more difficult to develop than the Transit system. It requires for example, a clock system several hundred times more accurate than the Transit oscillators and as a result each Navstar has a precious cargo of three Rubidium atomic clocks.

In terms of actual applications, the Navstars contain two separate systems; one for civilian users and a more accurate system for the military. Civilian users in all kinds of ocean going vessels can achieve accuracies of 300 ft (100 m) — as contrasted with the military's 30 ft (10 m) in all three dimensions and 0.1 ft per second (0.03 mps) in velocity. So accurate are the Rubidium atomic clocks that certain effects predicted by Einstein's General Theory of Relativity have to be taken into account when receiving signals from the Navstars.[33]

Successful Navstar tests have been made aboard aircraft, helicopters, ships, surface vehicles and even by individuals on foot. A 25-lb (11.3-kg) receiver now allows individual use. To supplement their space, missile, and general navigational applications, the Navstars facilitate such diverse military tasks as precise weapons delivery (ie, accurate missile attack), aircraft runway approaches, mapping, aerial rendezvous and refuelling, air traffic control, and rescue operations. By 1985, the full 18 satellite networks will be in place and some 19 000 users are anticipated.

In October 1982, the Soviets launched the first three GLONASS (Global Navigation Satellites System) navigation satellites. They were Cosmos 1413, 1414 and 1415. The GLONASS uses exactly the same methods and even the same orbit as the US Navstars. (One writer has affectionately dubbed them 'Navstarsky'). The launch vehicle used was the D-1-e.

Military communications

A major factor in the conduct of a battle is communication. Delayed reports, failure to receive new orders and confusion about the status of combat units all go to make up the fog of war. It comes as no surprise then that the prospect of improved communications spurred early military interest in satellites.

The US military communications satellite program explored various avenues, including the Army's delayed-repeater satellite, named Courier, and the Air Force's controversial West Ford Project. The latter was a passive communication system involving 400 million human-hair-size, copper dipole antennas in a ring 2000 miles (3218 km) above the Earth, which the astronomer Sir Bernard Lovell criticized as 'space pollution', causing a major furore at the time although the constituents actually dispersed without lasting or dangerous effects.

The first operational military communications satellite was the Defense Satellite Communications System I (DSCS-I) which employed small satellites in sub-synchronous orbit. They were launched eight at a time aboard Titan IIIC research and development vehicles. As their orbit was slightly lower than geosynchronous altitude, an observer would see them drift slowly across the sky.

Today the US uses several types of military communications satellites for different purposes. The mainstay of US strategic communications at present is the DSCS-II which carries two transponders capable of providing 1300 voice-channels, or 100 million bits per second of digital data. The DSCS-II embodies nearly ten years of steady improvement in satellite technology. This includes geosynchronous orbit, narrow beam and Earth-coverage antennas, secure communications and telemetry and high-capacity transponders for use with both small mobile and large fixed antennas. Another advanced feature is that the satellites can be shifted in geosynchronous orbit to new operating locations. They are launched in pairs aboard a Titan IIIC, the first having taken place in November 1971 with the first series completed in December 1978. Eight were successfully placed in geosynchronous orbit. Although early DSCS-II satellites suffered technical problems that limited their usefulness these deficiencies were soon corrected.

Their successor is the DSCS-III — a larger, more capable satellite placed in geosynchronous orbit by a Titan 34D or the Space Shuttle. Designed to last ten years, the DSCS-III has six super-high-frequency channels transmitting through three antennas. The complete program will consist of six satellites, two of them spares. The first was launched in October 1982 by a Titan 34D, and was paired with a DSCS-II. This was also the maiden flight of the Titan 34D launch vehicle.[34] The DSCS series is principally concerned with the transmission of orders and data from one major facility to another.

The use of satellites for tactical communications has long been the subject of research and actual engineering development work began with the Lincoln Experimental Satellites and TACSAT 1. The system now operational is the new Fleet Satellite Communications Program (FLTSATCOM). As the name suggests, it is primarily for use by the Navy, which funds the program. In the course of development, however, the Air Force realized that FLTSATCOM could also meet its requirements for global communications with strategic aircraft.

Using a dish-shaped and a separate vertical antenna, the FLTSATCOM has a capability of 23 channels. The Navy uses a fleet-broadcast channel which can accommodate 16 teletype signals and 9 fleet-relay channels, as compared with the Air

The NATO-III communications satellite uses a wide-beam transmit antenna for coverage from the US to the Middle East and a narrow-beam antenna for European use. Geostationary orbit allows continuous use and satellites of this type now provide 80 per cent of all military communications.

Force which uses 12 narrow-band channels and a wide-band channel. The Navy employs FLTSATCOM for worldwide, high-priority, ultra-high-frequency communications with aircraft, ships, submarines, and ground stations. The Air Force draws upon it for communications with SAC aircraft. At the last count, 4 C-135 airborne command posts, 20 B-52s and the SAC headquarters had been equipped with terminals. The first FLTSATCOM was launched into a geosynchronous orbit in February 1978 by an Atlas-Centaur.[35]

Quite distinct from the US military communications program but utilizing its technology is the NATO III communications satellite program. Intended to function as part of NATO's communications nexus, the satellite was expressly designed for capacity, versatility and reliable performance.

NATO III is used for both strategic and tactical communications and possesses two antenna beams. A wide beam links the European NATO countries with the US and Eastern Canada, while the separate narrow beam for use by small terminals such as ships in the Atlantic or mobile units covers Europe. It has three channels for voice, data and teletype, and its three constituent satellites were launched in April 1976, January 1977 and November 1978 respectively; the second and third NATO IIIs serving as on-orbit spares. Thor-Delta launch vehicles placed them into standard geosynchronous orbit.[36]

The riddle of the SDS

The final Western communications satellite is an enigma. Known simply (and tantalizingly) as the Satellite Data System (SDS), no pictures have yet been released of it. When it is launched by a Titan IIIB-Agena, its orbital elements are often withheld. The first SDS was launched in August 1973, the published orbit being highly elliptical: 24 434 by 236 miles (39 315 by 380 km), inclination 63°.[37]

The satellite's purpose is equally obscure. Like

FLTSATCOM, it carries equipment to transmit messages to SAC (Strategic Air Command) aircraft in the polar region. (In the case of FLTSATCOM this is because the satellite is located over the equator and communications with it are poor in the higher latitudes.) It is very likely that the SDS relays digital-imagery signals from the Key Hole-11 reconnaissance satellite. These signals would be transmitted from the KH-11 up to the SDS and then down on to ground stations, and relayed finally to the National Photographic Interpretation Center.

Soviet communications satellites fall into three categories. The Soviet military is believed to use the Molniya 1 for real time (or simultaneous) transmission of military telephone, television, telegraph and facsimile messages. First launched in April 1965 by an A-2e booster, the Molniya 1 satellite assumed a 23 357 by 310-mile (40 800 by 500-km) orbit, inclined at 62.9°, with a period of 12 hours.

This orbit was chosen because of difficulties posed by the Soviet Union's northerly location, and also because placing a satellite in a geosynchronous orbit is complicated and calls for a reduced payload. With a Molniya orbit, approximately eight hours of coverage can be achieved from a single satellite each day. When the Soviets began flying early warning satellites, they employed an orbit very much along these lines.

The Molniya 1 series carries three transmitter systems and two separate dish antennas. More advanced versions of the Molniya series have since appeared. Molniya 2 handles Soviet domestic programs and its frequencies comply with those of the International Intelsat system; Molniya 3 is used for color television and the US-USSR hot line. Molniya 1 flights continue to be made, which points to military usage. Fixed and mobile ground stations provide communications with the Molniya satellites.[38]

The USSR began to fly geosynchronous-orbit satellites in 1974 and there has been a slow shift to employing these more. The bandwidth of the satellites' transmitters suggests that they will be used for relay of data from digital imagery satellites at some future time.[39]

Soviet tactical communications are handled by a series of satellites with a unique launch profile. These 'storage dump' satellites are launched 8 at a time by C-1 rockets into 932-mile (1500-km) high circular orbits, 2 or 3 times per year; 24 to 30 of them are operational at any one time.[40] They appear to provide real time (or simultaneous) communications over the USSR and Eastern Europe for military command and control. On a worldwide basis, they could provide multiple channels for delayed messages which are important but do not require real time coverage. The data is radioed up to the satellites, recorded, then played back later.[41]

Soviet military doctrine allows little or no initiative below the regimental or divisional level. This is in keeping with the

ATLAS CENTAUR

poor view of initiative and spontaneity in the Soviet Union as a whole.[42] To carry out such strict control of military forces, however, requires multiple reliable lines of communications, and may help to explain why the Soviets maintain such a large network of satellite communications.

The final Soviet communications satellite is rather more obscure. Launched singly by a C-1 into a 512-mile (825-km) circular orbit once or twice a year, only one is active at a time. They do not fit into the other orbital networks; they lack the height which is crucial for real time coverage; neither do they transmit recognizable signals in the manner of navigation satellites. It has been suggested by some sources that they are a 'storage dump' for the transmission of information from secret agents and remote sensors located in various parts of the world.[43]

According to published reports, the Soviets have placed special detection equipment around certain US military bases and missile test facilities — equipment which broadcasts to satellites.[44] Soviet agents are thought to have special radio devices capable of receiving messages via satellite. The coded messages are compressed into a second, so that they sound like static to the casual listener. The receiving unit records the message and slows it down so the agent can decode it.[45] There may well be US equivalents to this system.

Satellites as front-line targets

This gallery of satellites represents the most advanced technology that the nations who constructed them can offer. Even to the unscientific eye, their brilliant blue solar panels and multifaceted surfaces display a peculiar beauty. This cannot however disguise the fact that in time of war or international crisis they become prime targets. Their crucial role in reconnaissance, early warning and military activities singles them out as immediate priorities for destruction in a large-scale conflict.

It is therefore no great surprise to learn that anti-satellite systems or ASATs have gone through almost as many generations as the satellites themselves. Nor is it particularly surprising, in the light of this, that the USSR has today on standby an ASAT system which may come into play at a moment's notice. The US — having already discarded two ASAT systems — is now about to introduce another. New work is being directed by both parties towards eliminating communications and advance warning satellites in high orbits, though this may take some time to achieve.

The real surprise lies in the history of these predatory ASATs which were not originally designed to attack mere reconnaissance satellites. They evolved as a countermeasure to something inestimably worse — orbital nuclear weapons. To this grim topic we must now turn.

Chapter 2

THE NUCLEAR FACTOR

'(…) the unleashed power of the atom has changed everything save our modes of thinking and we thus drift toward unparalleled catastrophes.'

Albert Einstein

During the Sixties, the phrase 'hostile satellite' was a sure reference to orbital nuclear weapons. Although they were never put into orbit, the threat of such weapons cast a long shadow of fear over the decade which influenced the actions of both the US and the USSR. By the end of the decade, both nations were able to spin networks of satellites, each satellite carrying a thermonuclear warhead and with it the imminent prospect of a conflagration that would utterly consume the Earth.

The military history of the Sixties is to a considerable extent the sinister story of nuclear orbital weapons. Their effect on the mass psychology of the decade was profound. It was a decade which saw the Berlin airlift and the confrontation between Khrushchev and Kennedy at the Bay of Pigs, and was dominated by the central fact that orbital nuclear bombs were under construction and were actually being exploded in space experimentally. Current space legislation proscribing such tests arose directly in reaction to them, and is a measure of their far-reaching effects.

To see in perspective the weapons race which culminated in the political threats and crises of the 1960s, it is first necessary to survey the origins of the technology involved.

Origins of orbital weapons

Space weapons have roots before World War II, pre-dating development of the atom bomb. In 1921 Hermann Oberth wrote about the possibility of erecting a space mirror. A rotating wire net covered with metal mirrors would be used to focus sunlight, reflecting it onto a limited spot or diffusing it for weather control. Oberth suggested, for example, that it might be used to keep northern shipping lanes and ports open, cause rain to fall on drought-stricken areas or to break up storms.[1] This gave rise to a rumour in 1945 that the Germans were in the process of developing a space mirror as a 'super weapon' which would concentrate the Sun's rays on an earthbound target. There were no such plans. In any case, most of its energy would have been dissipated by the Earth's atmosphere. However, use of such a mirror for climate modification is still considered valid.[2]

In 1948, Dr Walter Dornberger, the former German Army general in charge of V-2 development, suggested orbital atomic bombs as a prospective future weapon.[3] This was only one of a number of space concepts to emerge in the decade which immediately followed the war. The first satellite proposals were made during these years and manned space flight and space stations were also discussed.

In terms of real development these years were barren ones in the US. The high cost of even the simplest satellite program was unacceptable and the usefulness of satellites too speculative for a serious hearing to be granted. Not until 1955, for example, was the scientific satellite Vanguard approved even as a low priority effort.

The central problem in the early post-war years was the lack of a long-range rocket program. Without such rockets, any space effort was stillborn. In the late 1940s, the decision not to pursue this course had made apparent sense. True, the V-2 had demonstrated that a large rocket was possible despite serious shortcomings in range, accuracy and reliability, but even with major increases in thrust and accuracy, the 1940s vintage ICBM would have been hard pressed to hit a target the size of a city. A manned bomber, on the other hand, could attack a small individual target such as an airfield or industrial complex with an accuracy of a few hundred yards, then fight its way back to base, re-arm and make another strike.

The large size of early atomic weapons was another drawback. To carry one, the ICBM would be of unmanageable proportions and the cost of its development insupportable in the austere post-war financial environment. The US military could not even maintain its ongoing programs in such a severe budgetary environment. There seemed no reason to invest heavily in large rockets.[4]

General 'Hap' Arnold, Dr Vannevar Bush, Dr Theodore Von Karman, Dr Hugh L Dryden and the members of the Army Air Force Scientific Advisory group all believed it was unwise to begin work on long-range missiles.[5] Dr Bush, Director of US scientific research during World War II, made a famous remark about ICBMs during a Congressional appearance in December 1945. 'In my opinion', he said, 'such a thing is impossible (. . .) I say technically, I don't think anybody in the world knows how to do such a thing and I feel confident it will not be done for a very long period of time to come.' It was decided to concentrate on manned bombers such as the B-52 and large unmanned cruise missiles for future long-range strike forces. Work on rockets stopped in 1947.

Orbital nuclear weapons were brought back to public attention with the publication of *Across the Space Frontier* in September 1952. It was based on a series of articles appearing in *Collier's Magazine* which covered the technical, medical and legal aspects of the imminent space age. It described a ten-year program costing some $4 billion to develop a three-stage Space Shuttle and build a large, permanent space station in polar orbit.

The military potential of this space station was not overlooked. When the book was published at the height of the Cold War, Stalin was in power, the Korean war had entered its second year and both the US and USSR were preparing to test their first hydrogen bombs. The station, it was suggested, could have access to atomic warheads in small winged rockets carried aboard a sub-station. This sub-station would be in the same orbit as the space station but 2400 miles (3861 km) behind it.

These space bombs might be jettisoned by means of a re-entry burn halfway around the world from the target, the bomb pulling ahead of the sub-station as it descended. As the atomic warhead re-entered the Earth's atmosphere, the space station would have line-of-sight control over it. The warhead would next transmit its speed and altitude to the station, corrections of the bomb's flight path being calculated and radioed down. Finally, the radar tracks of bomb and target would be aligned by the guiding space station and the atomic bomb exploded. The authors considered such a station capable of dominating the Earth.

A counterattack on the station by unmanned rockets was dismissed in some quarters as impractical and the station would anyway carry missiles for self-defense in the event that it was attacked by a manned vehicle.[6] Some believed that on the contrary the station would make an easy target for a guided missile launched from the ground. They contended that such an interceptor would be simpler to construct than the shuttle used to supply the space station.[7] What emerges very clearly from the controversy is that even at this early stage the space bomb and ASATs were associated.

Fundamentals of an orbital nuclear bomb Such were the speculations which kindled fears in the popular imagination throughout the 1960s. But speculation aside, what would an orbital nuclear weapon actually require in technological terms and how close to achieving these requirements were the nations involved?

An orbital nuclear weapon requires several elements to function. It would probably have a high yield; 10 to 20 megatons would be a reasonable estimate. To illustrate this point it should be noted that the hydrogen bomb has an explosive yield measured in *millions* of tons of TNT (megatons), while the atomic bomb is measured in thousands (kilotons); and the Hiroshima atomic bomb had a 20-kiloton yield.

In other ways it would resemble an ICBM warhead. It requires heat protection during its descent and would carry a fusing system for detonation. The warhead would be attached to a module carrying the support equipment, the primary function of which is to make the weapon re-enter the atmosphere to attack on command.

A space bomb cannot simply be dropped; once detached from a satellite it would continue in the same orbit. For re-entry to occur it has to be slowed down by a retro-rocket. An attitude control system is needed to orientate the weapon. This entails small rocket thrusters, propellent tanks and associated plumbing and the module would also contain the guidance package of horizon, Sun and star sensors, gyros and computers.

The weapon's accuracy would of course depend on how well the guidance system operated. The retro-burn would have to be timed to the split second; fractionally too early or too late and it would miss. Accurate orientation is another requirement. The retros must fire at the correct angle. If not, the weapon either re-enters too steeply and burns up or is pushed into a higher orbit. Power for these systems might come from solar cells because unlike an ICBM which can draw power from internal batteries during its 30-minute flight, an orbital nuclear weapon would need a functional lifetime of many weeks or months. Finally, a communications system is called for to arm the weapon and set in motion the attack sequence. Obviously, it would need the highest possible reliability and multiple safeguards.[8]

What were the possible approach paths? A circular polar orbit was one possibility, in which case the weapon would pass over all of the Earth's surface twice each day. Another suggestion was to place the weapon in an equatorial orbit that would not overfly the target country. In wartime, its orbit would be shifted by an on-board engine (such an engine requires a large fuel supply, cutting into the available payload). The weapon could assume a very high permanent orbit or even be put into deep space.

Various deployment options presented themselves. A small number of weapons could be orbited, their presence either kept secret or revealed for blackmail purposes. Or an effective force of orbital weapons — representing a significant percentage of a nation's striking power — might be launched during a developing crisis or as a continuous deterrent patrol (similar to the airborne alert B-52s or Polaris submarines).[9] The goal in either case is to complicate the defensive efforts of an opponent, with multiple threats from different quarters.

Right: The thermonuclear breakthrough ushered in the ICBM race. Von Braun's original V-2 is shown alongside Atlas, the first US ICBM project. The Thor IRBM, ready in 1956, is dwarfed by the Soviet SS-6, which was designed before the breakthrough. The SS-6 remained unsurpassed in size until the 165-ft (50-m) Saturn I flew in 1961

This does not exhaust the nightmarish possibilities. Another suggestion was to put the bomb into a highly elliptical orbit so that it would not return to Earth until perhaps 48 hours after launch. The attacking nation could thus use it to force concessions from an adversary during a crisis. If the adversary capitulated within the allotted time, the weapon would be exploded in deep space by radio command.

An ordinary ICBM, in similar fashion, could place its warhead in a temporary orbit. This might take place at the first sign of an enemy ICBM attack to avoid the possibility of being caught on the ground. Another option would be to use an orbit to extend the range of an ICBM to global distances. In both cases, the weapon spends only a brief time in orbit: from less than one revolution to a few hours.[10,11]

The latter route was in fact the one eventually chosen by the USSR.

The ICBM race gets underway

Time passed and by 1953 the ICBM had become practicable. The change was ushered in by the 'thermonuclear breakthrough', laboratory tests and theoretical studies having indicated that with hydrogen bombs a nuclear warhead could be greatly reduced in size. This meant that ICBM engines, airframe, plumbing and controls shrank in their turn. Both the development and unit costs were now attainable.

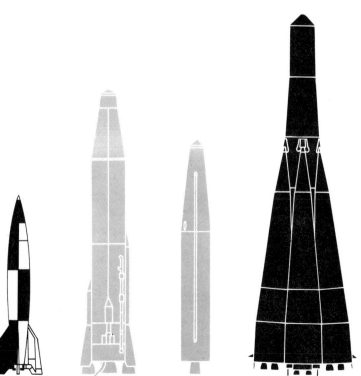
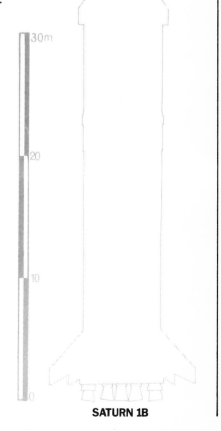

V-2   ATLAS D   THOR   SS-6   SATURN 1B

The vastly increased destructive power of the hydrogen bomb, coupled with its small size, also reduced the need for missile accuracy. A thermonuclear warhead had only to strike in the general vicinity of a target to destroy it completely. The ICBM was a very real possibility as a result and the US Atlas ICBM program began.[12]

The Soviets were less inhibited by cost effectiveness studies than the US. Since the war's end, an extensive Soviet rocketry program had been maintained and during the last years of Stalin's life preliminary design work was underway on the big SS-6 ICBM. Static engine tests were made and during the summer of 1955 work began at the Soviet missile test center near the small rail stop of Tyuratam. Because the SS-6 was designed before the thermonuclear breakthrough, it was truly huge compared to US ICBMs and not until Saturn I was flown would it be surpassed as the world's largest rocket.[13] It was to all intents and purposes a 1940s ICBM. The Atlas in the US and SS-6 in the USSR were being readied for their first test flights early in 1957. The first Atlas launch was made on June 11 — it blew up at 10 000 ft (3048 m).

Meanwhile, at Tyuratam, the SS-6 was making stage separation and maximum altitude flights. Beginning in May and lasting until August, the SS-6 made long-range test flights. Maximum range was 3500 nautical miles (6482 km), meaning that from the northwest USSR, the SS-6 could hit the northeast part of the United States: Maine, Boston and New York City. The northern part of Canada and all of Alaska were also in range and all of Europe as far as the Azores and the Canary Islands. From the Plesetsk site — the only location at which the SS-6 was operationally deployed — it could hit targets vital to Western security. On August 26, 1957, the Soviets announced that they had an ICBM.[14]

The Eisenhower Administration adopted a calm public stance, playing down the importance of the Soviet test. An Administration spokesman said the missile was not even a prototype and that there would be no significant difference in deployment time between US and Soviet weapons.[15]

Six weeks later, on Friday, October 4, 1957 an SS-6 put Sputnik I into orbit, bringing to the world's attention Soviet accomplishments in rocketry. The words 'space race' and 'missile gap' entered the US vocabulary. ICBM development became a breakneck race.

The development of large rockets also meant that the highway to space was open. With upper stages, payloads of several tons were possible, introducing the specter of orbital nuclear weapons. Studies were already in progress and covered a wide area from low Earth orbit to the Moon.

Research into the feasibility of interplanetary strategic space missions was also conducted in 1959 and 1960, and included the use of the Moon as a missile launching base.[16] The advantage here, according to the concept's supporters,

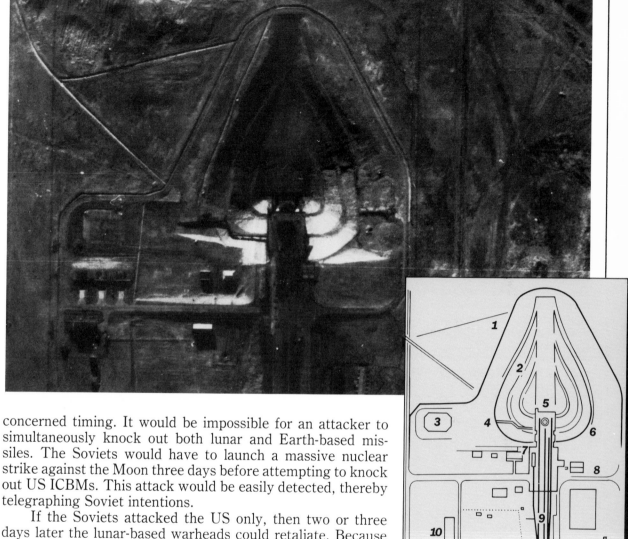

concerned timing. It would be impossible for an attacker to simultaneously knock out both lunar and Earth-based missiles. The Soviets would have to launch a massive nuclear strike against the Moon three days before attempting to knock out US ICBMs. This attack would be easily detected, thereby telegraphing Soviet intentions.

If the Soviets attacked the US only, then two or three days later the lunar-based warheads could retaliate. Because of the lower gravity on the Moon, smaller boosters would be required and as one side of the Moon always faces the Earth guidance of the warheads would be simplified.[17] Absurd as it seems, the principal objection to this scheme was the sheer cost of transporting the missiles, crews and equipment to the Moon and maintaining the base.

The Boeing Aircraft Company

The Boeing Aircraft Company researched orbital weapons systems extensively in the early 1960s, and was at work on the X-20 Dyna-Soar (a dynamic-soaring plane) — a small winged space glider that carried a pilot. It was to be used for research into winged re-entry from space and aerodynamic data at speeds greater than Mach 6, with a view to providing a system adaptable to military purposes. The

The SS-6 ICBM launch pad at Tyuratam as revealed by a U-2 photograph released by the CIA. A civil engineering study of the pad and buildings might indicate the feasibility of full-scale development of the missile.

1 *Rain ditch*
2 *Flame pit*
3 *Block house*
4 *Block house passageway*
5 *Launch pad*
6 *Tower*
7 *Railcar covered shelter*
8 *Causeway*
9 *Roadway*
10 *Support buildings*

Above: The Dyna-Soar crew beside the original mock-up. Dyna-Soar applications for various military missions were investigated in the early 1960s.
Right: Last of the lifting bodies to fly in 1975, the X-24B touches down at Edwards AFB, California.

Dyna-Soar was a notable forerunner of the Space Shuttle and had a payload of 1000 lb (453 kg) and a compartment of 75 cubic ft (2.12 cubic m). About the size of the B-26 bomb bay, this payload compartment was fitted with air conditioning and various equipment to enable the vehicle to carry military or scientific cargos.[18]

Boeing investigated a whole family of space bombers. They were Dyna-Soar II, Dyna-Soar III and Dyna-Mows (the latter meaning Manned Orbital Weapons System). The studies explored modifications to the Dyna-Soar vehicle, suitability for bombing and re-supply missions, and identified areas for further research. Dyna-Soar applications for a variety of military missions were studied right up to the program's cancellation in December 1963.

Boeing had meanwhile completed an internal study called Boss-IOC (or Bomb Orbital Strategic System — Initial Operating Capability). The study concerned an unmanned vehicle with ballistic re-entry, the intention being to place the weapon in an eliptical orbit with a 24-hour period. This in turn expanded into Boss-Wedge (Weapon Development Glide Entry). Boss-Wedge determined the characteristics of second generation vehicles with a wide variety of military tasks (both combat and non-combat missions) in mind. It included a study of hardware and subsystems for the Air Force Special Weapons Center. Undertaken in late 1960, the latter concentrated specifically on boosters and components likely to be available after 1970.

The concept was taken further under a Wright Air Development Division contract. Designated SR-79821, this study (which ran from October 1960 to April 1961) was also known as the Advanced Earth Orbital Weapon System (Low Orbit), and looked at weapons-carrying satellites orbiting at altitudes up to 1000 nautical miles (1852 km) from Earth.

Blueprints for orbital nuclear weapons were becoming ever more sophisticated and feasible. This particular program explored the strengths and weaknesses of three different space bombers, with and without pilots, employing both ballistic and winged re-entry (the latter course anticipating the Space Shuttle). A later study envisioned an armed space glider capable of maneuvering, surveillance, orbital bombing and damage assessment.[19]

Soon afterwards Boeing was contracted by the Air Force Ballistic Missile Division to conduct yet another study, known this time as the Advanced Earth Orbital Weapon System (High Orbit). It concentrated over six months on vehicles with orbits above 10 000 nautical miles (18 520 km) and approximately three-quarters of the study dealt with the Boss-IOC type vehicle in a 24-hour orbit.[20] If orbital bombs were to be manufactured as a matter of policy, it now remained only to iron out minor details.

The charred ablative covering on this X-23 lifting body testifies to the effects of re-entry. The thermal tiles of the Space Shuttle are a subsequent refinement.

The recallable ICBM

Also explored was one important system which tried to circumvent the terrible finality of ICBM warfare. The ICBM is unrecallable which obviously constitutes a major limitation. In practice, this means that if warning of Soviet attack is received, the US SAC bombers take off and head for their targets. If the warning is genuine and Soviet missiles are incoming, the bombers (themselves prime targets by this time) strike as the vanguard of the US counterattack. Meanwhile the US ICBMs must wait out the first wave of incoming Soviet ICBMs, with the attendant risk that they may be destroyed.

Given the possibility of successfully carrying out a partially pre-emptive strike in this way, some US military circles felt this made the prospect of a nuclear attack more likely. A recallable ICBM would remove this possibility, since such an ICBM would be fired on the first warning and neutralized before it struck the target if the alarm was false.

The actual proposal was that a Titan II be fitted with a 3-ton armed satellite. Upon warning of an attack, several hundred such missiles would be launched into nominal orbits at 100 nautical miles (185 km) altitude. There would be an 18-minute period of line-of-sight communication each time the satellite passed over its ground station, during which it might receive the re-entry (ie, attack) command. From orbit, it could attack a target from a distance of about 5000 nautical miles (9260 km) out as far as 16 000 nautical miles (29 632 km). The weapon was recallable insofar as if during the 18-minute command interval the final attack order was not received the vehicle would automatically discontinue attack preparations, disarming itself and splashing down in a remote ocean area.[21] One of a number of advanced weapons systems, this so-called Positive Control Bombardment System was to be launched by the Titan II and formed part of the Air Force's ambitious space plans.[22,23] During fiscal year 1963, $1.5 million was budgeted for the study of orbital space weapons.[24]

Orbital nuclear weapons were now highly controversial. A fundamental question dominated debate: would a US orbital weapons program provide a deterrent or retaliatory capability to justify the cost and risks? Supporters of such weapons pointed to their mobility and direct psychological effects. Moving at orbital velocity, a space bomb travels over most of the world. Supporters considered that this would make them both easily discernible and much less vulnerable to attack than a fixed silo. Missiles in their silos and Polaris submarines, on the other hand, were out of sight and out of mind.

Opponents of orbital nuclear weapons argued very differently. They considered that the weapons *were* vulnerable to attack. After all, the position and orbit of an armed satellite could be determined by a tracking network and thereby attacked. A primitive ASAT was even now at hand and armed satellites could be destroyed in short order.

The fact was that, given the technology then available, an orbital weapon was less accurate than an ICBM. A fixed ICBM silo can be located very precisely, within a few tens of feet. An orbiting satellite moves through space and its ever changing position in three dimensions and corresponding velocity cannot be determined so precisely.

What is more, an orbital bomb's gyro system is another drawback. ICBM gyros operate for only a few minutes during powered flight. The gyros of an orbital bomb have to operate for much longer periods. Like all mechanical devices, they are prone to small errors. These errors are not serious during the short operating lifetime of an ICBM, but in an orbital nuclear weapon with its longer lifetime they accumulate alarmingly. As gyros orientate the satellite for retro-fire, pointing errors in this firing would make the warhead miss its target by a wide margin. Inaccuracy on this scale limits a space bomb to a large-scale target — a city, for instance.

Nor were these by any means the only problems. An ICBM in its silo is constantly monitored and maintained. Once in orbit, a space bomb is impossible to maintain or repair, so that it is unreliable on this score as well. It was calculated that to match any given force of ICBMs five or even ten times as many orbital weapons were needed.

Expense was a crucial factor, as always. An extensive communications system is needed to control space bombs and the large boosters needed to orbit them. An orbital weapons network would be up to possibly 20 times as expensive as an equivalent force of ICBMs, making the cost of a space bomb program somewhere between $100 and 200 billion.

The terrible risks of orbital weapons   Fearsome deficiencies existed in the areas of security and safety. With hundreds of space bombs overhead, there was the very real danger that a stray radio signal might bring one down by accident. Another risk was that a madman might intentionally trigger an orbital nuclear weapon to start a conflict.

The unnerving possibilities did not stop here, it was argued. A booster failure during launch would cause the warhead to fall back to Earth with cataclysmic consequences. If the weapon was placed in a faulty orbit, decay and an uncontrolled re-entry were likely. After all, launch failures of unarmed missiles were depressingly familiar during the early 1960s. The belief of the space bomb critics was that ICBMs in hardened silos and Polaris submarines hidden within the oceans provided a cheaper and more efficient deterrent than orbital nuclear weapons.[25]

The Positive Control Bombardment System concept, although it avoided many of the deficiencies noted, still ran into opposition. It was pointed out that the sudden launching of several hundred missiles inside a few minutes would register on enemy radar as a US attack. This might actually trigger a Soviet counterattack rather than deterring it.[26] The Eisenhower Administration was dedicated to a space for peace policy, an attitude that effectively ruled out military space programs. The thought of cluttering up the heavens with bombs was anathema in anyone's book.

Spokesmen stressed that the US had decided to unilaterally abstain from the development of orbital bombs.[27] Deputy Defense Secretary Roswell Gilpatric made the definitive statement in a policy speech read and approved by President Kennedy:

'The United States believes that it is highly desirable for its own security and for the security of the world that the arms race should not be extended into outer space, and we are seeking in every feasible way to achieve that

Above and opposite: This photograph of Hiroshima was taken close to ground zero. The area of devastation is represented on the corresponding diagram by the innermost circle. This circle is the area subjected to a blast 'overpressure' of 20 lb/sq in (9 kg/sq cm). The other circles are the projected areas of devastation for other weapon strengths. The outermost one shows the destructive radius of the 100-megaton space bomb of which Khrushchev boasted.

purpose. Today there is no doubt that either the United States or the Soviet Union could place thermonuclear weapons in orbit, but such an action is just not a rational military strategy for either side for the foreseeable future. We have no program to place any weapons of mass destruction into orbit. An arms race in space will not contribute to our security. I can think of no greater stimulus for a Soviet thermonuclear arms effort in space than a United States commitment to such a program. This we will not do.'[28]

It was felt within the US Government that the Soviets would find orbital nuclear weapons unattractive for the same reasons.

Orbital nuclear weapons and the USSR

Many of the arguments against orbital nuclear weapons resembled those against ICBMs a decade earlier: namely, that they were inaccurate, unreliable, expensive, and better substitutes existed. Yet these drawbacks had not deterred the Soviets from embarking on a development program in the case of ICBMs.

What were the conceivable advantages of orbital weapons from the Soviet point of view? First, such weapons would overcome the West's geographical advantage. The US had bases in Europe, North Africa and the Far East from which missiles and bombers could attack the USSR. In the late 1950s, the Soviets had no such equivalents. Any attack would have to be made from the Soviet Union itself; a considerably greater distance. Orbital nuclear weapons made distance an unimportant consideration.

The Soviets had demonstrated the basic technology necessary for a space weapons program by the early 1960s. The Vostok spacecraft could orbit a large payload for several days before landing at a pre-arranged time and place.[29] In two other areas as well — large nuclear weapons and large space boosters — the Soviets had significant leads. While the US were constructing a large number of relatively small nuclear warheads of a few hundred kilotons yield each, the Soviets, subscribing to a 'big is better' philosophy, were at work in the early 1960s on a superbomb program.[30]

Proof positive was supplied by a number of atmospheric tests. On October 30, 1961, the Soviets tested a 50-megaton device. A 30-megaton bomb test followed in 1962. Subsequent analysis indicated that the first device had been encased in lead to lower its yield; with a uranium casing, its yield might have been equivalent to 100 million tons of TNT.[31] A weapon of this type would totally destroy everything within a 12-mile (19-km) radius of ground zero. Fallout would be spread over an area equivalent to the states of Massachusetts, Connecticut, Rhode Island and New Jersey.[32]

These large bombs would be suitable for space weapons applications, their high yield lessening the accuracy problem and the number of weapons needed. Another possibility was to surround the weapon with cobalt; the resulting fallout would render vast areas uninhabitable.

Such developments caused Dr Donald Brennan, President of the Hudson Institute, a 'think tank' specializing in nuclear policy, to visualize an apocalyptic weapon. The Soviets, he said, might place a small number of very large weapons into low Earth orbit. These weapons, with yields of several hundred megatons, would not re-enter the atmosphere — they would be detonated in space instead. The thermal flash would literally fry broad areas of a continent. Only ashes would be left.[33]

If the 100-megaton bomb was to be an effective weapon, a means of delivery — in other words a very large missile — would be required. A 100-megaton space bomb could be expected to weigh about 40 000 lb (18 144 kg), roughly three times the orbital payload of the Soviet A-2 booster.[34] Even with the less stringent performance requirements of an ICBM, the size of a 100-megaton warhead demanded a much bigger rocket than any the Soviets had at the time. Thankfully, no such missile was ever deployed but some Western military observers held that the Soviet D booster was originally designed to carry the 100-megaton 'city buster' warhead.[35]

First flown in 1965, the D booster has been used to launch the Proton Cosmic Ray satellites, Soviet probes to the Moon, Venus and Mars, as well as the Salyut space stations.[36] It has

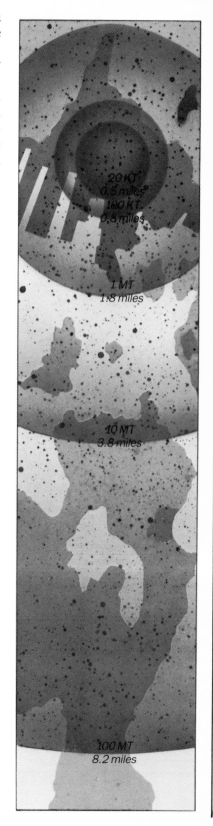

20 KT
0.5 miles

100 KT
0.8 miles

1 MT
1.8 miles

10 MT
3.8 miles

100 MT
8.2 miles

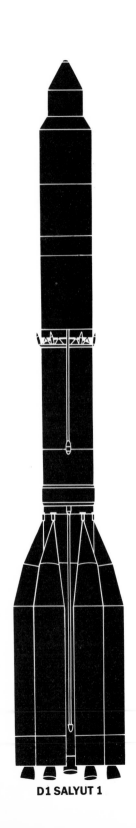

D1 SALYUT 1

a payload of 44 000 lb (20 000 kg) which accords with the space bomb estimates and its early development took place alongside the superbomb program. The strongest circumstantial evidence in support of the D booster's suspected military origin concerns its propellents: nitrogen tetroxide and asymmetrical dimethyl hydrazine. These are storable, so that a rocket employing them could be kept fueled indefinitely and launched on short notice. This suggests a military application for the D booster as opposed to a civilian one.[37] Others contend that despite this the D booster — like the US Saturn family of rockets — was designed from the start as a peaceful space launch vehicle.[38]

After examining Soviet capabilities, policies and statements, it was predicted in 1960 that the USSR could launch orbital nuclear weapons in the second half of the decade.[39] Aware of major deficiencies in the coverage and response time of its space tracking network, the US felt vulnerable. If the USSR fired space weapons, the US would not have the precise orbital data necessary in order to mount an ASAT counterattack.

This fact first came to public attention in January 1960 when technicians with the Navy's space surveillance system (SPASUR) discovered an unknown satellite in a polar orbit. It was in a 1074 by 134-mile (1728 by 215-km) orbit at an inclination of 79°; the period was 104.5 minutes. Dubbed the 'Dark Satellite' the object was believed to be tumbling and was slightly smaller than the US Discoverer reconnaissance satellite. Initially, it was believed to be of Soviet origin; either a reconnaissance satellite or the final stage of the Luna 3 Moon probe.

The 'Dark Satellite' was considered so secret that the House Space Committee was not told of it.[40,41] The mystery evaporated in late February when it was identified as the re-entry capsule from Discoverer 5, which had been launched in August 1959. A faulty retro-fire had given it a higher orbit.[42] The US found it distinctly disturbing that although the 'Dark Satellite' was first detected by the SPASUR network on December 19, several weeks elapsed before serious notice was taken of it.[43]

The Soviets launched Cosmos 1 two years later on March 16, 1962. It was placed into an orbit with a 49° inclination instead of the usual 65° and was not picked up by a US tracking station until four hours after launch.[44] Cosmos 2 also went undetected for two orbits.[45] Later that year, the dual flight of Vostok 3 and 4 gave new urgency to demands for improved tracking facilities.[46]

A Soviet orbital weapons program could not be pursued in isolation; it would constitute an integral part of Soviet military, domestic and foreign policy. Before the Cuban missile crisis, the Soviets had made various threats against the West. Nuclear missiles loomed behind these threats. Examples of this include Soviet relations with Britain and France during the Suez crisis; with Norway, Turkey and Pakistan over their involvement in the U-2 incident; and relations with the US in July 1960 over Cuba.[47,48]

The *New York Times,* in an editorial printed shortly after the Soviet ICBM announcement, referred to this practice as 'a war of propaganda terror' which gives some indication of the mood of the day. Other activities such as rocket firings, nuclear tests and space shots were scheduled in such a way as to put pressure on the West, or so it seemed. The 50-megaton nuclear test appeared on one level to be part of the Soviet effort over Berlin and the bid to achieve a German peace treaty.[49] The test itself was technically unnecessary according to some sources.[50,51] Meanwhile the Soviets were exhorting Western European nations to pull out of NATO and adopt neutralist attitudes.[52] Khrushchev sought relentlessly to embarrass and subvert Western interests while avoiding nuclear war.[53] His inclination to use pressure as an instrument of policy might make an orbital weapons program seem attractive, it was argued.

Sputnik was a case in point. Soviet technological success in this instance had a profound effect on the US. People stood at night in their backyards watching a small spot of light move silently overhead.[54] Sputnik was a symbolic triumph and it unnerved great numbers of people in the US.

Khrushchev's public 'missile rattling' Khrushchev used ballistic strength as a political weapon in this period, in a bid to obtain concessions from the West. Even though the Soviet ICBM force was a token one, it effectively neutralized the much more numerous US bombers. The Soviets' strategic forces were a minimum deterrent, able to inflict damage on the US but unable to destroy every target. Khrushchev exploited this psychological advantage in support of his aggressive foreign policy.[55] A key factor in all this was the ability to bluff while keeping the West in the dark about just how few missiles the USSR really had. With similar manipulations, even a small number of orbital weapons would provide a disproportionately powerful political impact.

Soviet spokesmen made occasional statements about space bombs during the late 1950s and early 1960s. Major General G I Pokvovsky in *Science and Technology in Contemporary War* published shortly after Sputnik I briefly alluded to the use of satellites for bombardment purposes. Khrushchev never tired of hinting at such things in the political arena.[56]

He addressed the Supreme Soviet in January 1960 and announced that he had 'in the hatching stage (. . .) a fantastic weapon'. As the Soviet space program gained momentum,

official statements became more explicit. On August 9, 1961 at a Kremlin reception for the Vostok 2 cosmonaut Gherman Titov, Khrushchev made direct public reference to orbital weapons. His remarks were aimed at the West when he said

'You do not have 50 or 100-megaton bombs, we have bombs more powerful than 100 megatons. We placed Gagarin and Titov in space and we can replace them with other loads that can be directed to any place on Earth.'[57]

A few months later on March 15, 1962 he boasted that 'we can launch missiles not only over the North Pole, but in the opposite direction, too (. . .) Global rockets can fly from the oceans or other directions where warning facilities cannot be installed. Given global missiles, the warning system in general has lost its importance. Global missiles cannot be spotted in time to prepare any measures against them'.

A week later an article by Lt Col V Larionov in *Krasnava Zvezda (Red Star)* declared that 'it is recognized in military strategy that outer space weapons will become the primary means for resolving strategic tasks since their operation cannot be linked with any concrete land, sea or air theater of military operation'.[58] Marshall Sergei Biryusov, Commander of the Strategic Rocket Forces, claimed in an article for Soviet Armed Forces Day less than a year later: 'it has now (. . .) become possible at a command from Earth to launch rockets from satellites at any desired time and at any point of the satellite's trajectory'.[59]

Such statements signified interest in space weapons at the highest levels of the Soviet Government. Khrushchev on the other hand seemed at times to be oblivious to the implications and consequences of his own remarks.[60] Headstrong and impulsive, he seemed occasionally to think that the announcement of a grandiose scheme meant its achievement was at hand.[61] Piercing the political rhetoric, it appears that the Soviets made initial studies of orbital nuclear weapons in the late 1950s and early 1960s. By the time of Khrushchev's statement at the Titov reception, they had probably weighed up the options, technical requirements and costs.

The technical limitations of orbital weapons did not perhaps deter the Soviets as they had the US. To judge from his comments, Khrushchev himself never grasped the subtleties of nuclear policy or military efficiency.[62]

The 100-megaton bomb of which he boasted is a good example. Two ten-megaton weapons properly placed could cause almost as much damage.[63] Khrushchev, however, subscribed to the 'bigger is better' philosophy even in the moral minefield of nuclear arms policy. It is sobering to reflect that what perhaps prevented a major orbital weapons program was the sheer expense. Committed to domestic economic development in an attempt to raise the Soviet consumer's lot above that of the US, Khrushchev could not afford an expensive space weapons program (and we have seen how such factors applied to the US as well). Khrushchev held that communism should be a philosophy of plenty offering a better life than under capitalism. With the growth of Soviet economic strength, he reasoned, the US position would be weakened as first the neutral countries and then the Western allies aligned themselves with the USSR.[64]

In Khrushchev's view, missiles offered a low-cost means of achieving nuclear parity with the West without the economic burden of large conventional forces. In 1955 and 1958, a total of two million men were demobilized from the Soviet armed forces. There seemed to be little reason for a large ground army in an age of nuclear bombs and ICBMs. Money and resources saved in this way could be applied to domestic economic development. Like many a Western statesman, Khrushchev modified his estimate of the international situation to suit domestic policies.[65] His pandering to consumers at the expense of the military clashed with defense interests and led to discontent amongst this powerful section of Soviet society.

Khrushchev's bluff is called When Khrushchev made his statement on space bombs in August 1961, the Soviets had considerable ability to control events by using their missile strength as a weapon of influence. In the next month, US intelligence — using photos from reconnaissance satellites — confirmed long-held suspicions. Khrushchev had been bluffing, and Soviet missile strength was only apparent. The US actually had the superior missile force.[66]

The first generation Soviet ICBM was inaccurate and extremely cumbersome. Faced with these obvious deficiences, Khrushchev decided to build only a handful of SS-6 missiles in 1960-61 rather than going ahead with full scale deployment. The lessons learned from the SS-6 contributed to two second generation missiles — the SS-7 and 8 which were deployed in underground silos. Their test flights began in 1961 and Khrushchev would have to wait several years before they were ready. In the meantime, the Soviets still sought to convey an image of overwhelming missile strength.[67] The effort was unconvincing and because of the decision to wait for the second generation ICBMs the Soviets now found themselves bringing up the rear of the missile race.

To add to Khrushchev's problems, Soviet economic performance, impressive during the 1950s, was deteriorating in the early 1960s. Bright promises about the coming age of abundance rang increasingly hollow. The struggling Soviet economy would be hard pushed to support a large ICBM crash program.[68] At such times, Khrushchev tended to take bold, aggressive and risky actions.[69]

His solution this time was to put medium-range missiles in Cuba, seeking at a stroke to upset the East-West military balance, force a solution to the Berlin question on Soviet terms and teach President Kennedy a lesson.[70,71] Kennedy's behaviour during the Bay of Pigs fiasco, at the Vienna Summit in June 1961 and his passive acceptance of the construction of the Berlin Wall emboldened Khrushchev. Gambling that Kennedy would vacillate long enough for the missiles stationed in Cuba to become operational, Khrushchev acted. His miscalculation brought the world to the brink of nuclear war.[72] Forced to retreat and conscious of the rapidly emerging ideological and military threat of China, he subsequently decided to establish better relations with the US. It was a decision which would greatly affect the subject of orbital nuclear weapons.

The Fractional Orbit Bombardment System

Robert S McNamara, US Defense Secretary during the Kennedy and Johnson administrations, held a press conference on November 3, 1967 at which he announced that the Soviets had been testing an orbital weapons system. This system, he claimed, might well be ready for deployment in 1968. The outer space treaty had been in effect just 24 days. The Soviet system was not unambiguously a bomb in space but rather an ICBM employing orbital flight to extend its range. Called the Fractional Orbit Bombardment System (FOBS), its booster was the SS-9 Mod 3 Scarp, also referred to as the F-1-r.

The vehicle had two stages plus the re-entry stage and its

SS-9   SS-10   F-1-r FOBS

Soviet SS-9 ICBM in a Red Square parade. This large two-stage rocket was used to launch the USSR's orbital weapon, the Fractional Orbit Bombardment System. (It was later adapted for the 'satellite killers' launched from Baikonur.)

total length was approximately 119 ft (36.4 m). The first stage, believed to burn nitric acid and kerosene, used 6 main engines and 4 verniers. The second stage was thought to have two main nozzles, possibly fueled by storable nitric acid and dimethylhydrazine.[73] The 'r' or re-entry stage, weighing about 9866 lb (4475 kg), comprised the retro-rocket and re-entry vehicle. The warhead was credited with a yield of 1 to 3 megatons (50 to 150 times the explosive yield of the first atomic bomb).

With the warhead placed in a low orbit, the ICBM had an unlimited range. It was able to approach the US from any direction, not just over the North Pole and consequently able to elude US north-facing early warning radars, thus allowing little or no warning. As the retro-rocket was fired about 500 miles (805 km) from the target, there would be only three minutes between retro-fire and impact, the 'r' stage being brought down before the completion of one orbit. Hence the term 'fractional' in its title.[74]

As the West saw it, the FOBS would act as a 'pathfinder' for a Soviet ICBM salvo attack, its main purpose being to destroy ABM radars. The FOBS launchings would be timed so that their warheads arrived before those of the ICBMs, suppressing US defenses and clearing the way for the following wave of missiles.[75] Other targets would include the US Command and Control network such as the White House, Pentagon and alternate command posts. This attack profile would disrupt the US retaliatory strike by preventing the relay of launch orders. The FOBS concept had serious limitations, however, first and foremost in terms of limited payload and accuracy. For this reason FOBS warheads were deemed ineffective against hard targets such as missile silos.

In defense of the new legislation, McNamara pointed out that the FOBS was not strictly speaking a violation of the space treaty since it did not complete a full orbit, nor did the

Two SS-10 ICBMs in Moscow's Red Square, November 7, 1968.

vehicles pass over the US or carry a nuclear warhead. A FOBS was simply an ICBM that followed an orbital rather than a ballistic path. McNamara was ignoring the fact that the two trajectories are separate and distinct. The view that an object is not in space unless it completes a full Earth orbit is technically and legally unsound, and his statements suggest an attempt to salvage the space treaty and to fend off demands that the US likewise construct a FOBS. McNamara's political and legislative acrobatics were successful nonetheless as the FOBS was sufficiently like an ICBM and sufficiently unlike the classic bomb in space to make the case ambiguous.

Despite this, it was widely held that even in an unarmed test form the FOBS constituted a violation in spirit of the space treaty.

The Soviet FOBS program had its roots in Khrushchev's missile speech of 1962. At that time, the Soviets were working on a third generation of ICBMs — the SS-9 Scarp and the SS-10 Scrag. These missiles were designed to have a range of 7500 miles (12 000 km) and to carry a 25-megaton warhead, superseding the existing Soviet ICBMs in both respects. With a reduced payload, they could form orbital weapons.[76]

US intelligence observed the large Soviet ICBM program during 1964, and reported that 66 SS-9 silos were under construction by the end of 1965.[77] The SS-10 was evidently designed as a fall-back system alongside the more technically advanced and therefore more accident-prone SS-9. SS-10 drew on cryogenic propellents, entailing protracted pre-launch fueling. Because the liquid oxygen which it used would boil away soon after fueling, launching could not be delayed. With the success of the SS-9 test program, the SS-10 was abandoned.

It had made its first public appearance during the 1965 May Day parade.[78] The Soviet radio announcer who described the missile claimed that it had an orbital weapons application. It appeared again during the November 7, 1965 parade to celebrate the Russian Revolution, and here too the SS-10 was referred to as an 'orbital rocket'.[79] Soon afterwards, the US State Department issued a public statement requesting the USSR to clarify their intentions regarding the UN resolution on space weapons and in December the Soviets gave private assurances that they were abiding by the resolution and would continue to do so.

Through *Pravda* the Soviets issued a counterblast, saying that the US 'by raising a racket about the Soviet orbital rocket, calculated to divert the attention of the world public from the American preparations in the cosmos. The activity of the USA (. . .) is subordinated to the idea of using space for military purposes (. . .)'[80] This was only one of a number of statements by the Soviets accusing the US of actual development of orbital nuclear weapons, extending the arms race into space, militarizing space and preparing for a new world war.[81]

The 'back door' attack. The low-orbit FOBS has unlimited range and goes undetected. By contrast, a conventional ICBM is bound to approach the US over the North Pole, activating the BMEWS defence screen.

FOBS leaves the drawing-board

A few days after the *Pravda* attack, the Soviets began what appeared to be the first flight test of FOBS hardware. The tests were long-range, sub-orbital firings. After the second-stage shutdown, the 'r' stage was observed to separate and retro-fire, falling on Soviet territory while the second stage went on to impact in the Pacific.

On September 17, 1966 the Soviets made their first orbital test of FOBS-related hardware. The flight used the F-1 booster based on the SS-9. The orbital inclination was 49.6°; a new specification. The second stage was left in a low elliptical orbit of 325 by 133 miles (523 by 214 km). A third stage went into an eccentric orbit of 628 by 174 miles (1010 by 280 km). The payload finally assumed a 650 by 101-mile (1046 by 163-km) orbit. It was intentionally exploded and a total of more than one hundred pieces were detected by US monitors.

The Soviets made no announcement of the launch, recalling by this the planetary failures of 1962 and 1963. Their silence again kindled Western interest. On November 2, 1966 there was a similar flight profile. Approximately 50 pieces of

An SS-9 sheds its cargo of multiple re-entry vehicles. The bright streak at the top of the photograph shows booster parts burning as they hit the atmosphere. The distinct streaks below it are made by the re-entry vehicles which would contain the nuclear payload.

debris were left in orbits ranging from about 310 up to 932 miles (500 to 1500 km).

Almost certainly FOBS related, the exact purpose of these two flights remains unclear. Did they suffer malfunctions? Were they perhaps research and development tests of the FOBS system before committing it to the operational one-orbit profile? Or did the missions test the FOBS in prolonged flight? The Soviets made no secret of the fact that it was possible to bring down orbital weapons during the first or subsequent orbits.

With the new year, they began an intensive FOBS test program. Perhaps because of the unwanted attention the unannounced launches had generated, these flights were given Cosmos numbers and their orbital elements were announced — except, that is, for the orbital period. This was never listed, indicating that the payload was brought down before the completion of the first orbit.[82]

Nine flights took place in all following a distinctive flight path. The first flight on January 25, 1967 was that of Cosmos 139 which entered a 130 by 89-mile (210 by 144-km) orbit. Next was Cosmos 160 launched on May 17. Two FOBS launches occurred in July, Cosmos 169 (July 17) and Cosmos 170 (July 31) to be followed by Cosmos 171 on August 8. Four more flights completed the test program: Cosmos 178 (September 19), Cosmos 179 (September 22), Cosmos 183 (October 18) and Cosmos 187 (October 28).[83]

At no time did the vehicle pass over the continental United States, although these test flights would indicate the attack accuracy to be expected.

Launch from Tyuratam occurred in the late afternoon or early evening. The typical orbit was 133 by 84 miles (215 by

Above: Boeing's X-20 Dyna-Soar was a manned vehicle designed for research into winged re-entry from space, and was therefore a forerunner of the Space Shuttle. Another study (BOSS-IOC) looked at an orbital weapon in the form of an unmanned vehicle with ballistic re-entry.

Left: On the alert for incoming orbital weapons and ICBMs. An artist's impression of the Pushkino anti-ballistic missile radar outside Moscow.

135 km) with an inclination of 49.6°. On each occasion three objects went into orbit: the second stage, the warhead, and an object referred to as a 'platform' which was possibly an inter-stage or a shroud. The warhead's orbital ground track passed over Siberia and Japan, then across the Central Pacific, along the eastern edge of South America, across the South Atlantic, over Africa and the Mediterranean. The 'warhead' finally passed over Soviet territory and before completing a full orbit, it retro-fired and re-entered. The flight path allowed Soviet ground stations to monitor the crucial de-orbit maneuver and re-entry.

The second stage, tumbling in orbit, decayed after a few hours to be followed by the re-entry of the more compact so-

Project Safeguard

One of MAD's consequences is that missiles are required to protect other missiles in their silos. This philosophy culminated in Project Safeguard. A $6 billion investment, Safeguard was constructed at Grand Forks Air Force Base, North Dakota to protect Minuteman ICBMs (see below). Astonishingly, it was closed down only a day after its official opening. As a result, the West now lacks any ABM defence.

Right: Safeguard depended on two kinds of ABM missiles: Spartan, which attacked at long range above the atmosphere, and the fast-reacting Sprint to catch warheads after they had entered the atmosphere. In both cases the 'kill' mechanism was nuclear. The two-stage Sprint had a very high acceleration and was launched from its underground cell by a gas-powered piston, the first stage igniting in mid-air. Rapid course changes after launch were brought about by a thrust-vector control system by fluid injection into a single propulsion nozzle, and then during second-stage flight by aerodynamic controls. Associated with Safeguard were Perimeter Acquisition Radars (PARS) located at strategic points around the United States. The Soviets have gone ahead with a similar system as an adjunct of an improved Galosh ABM for the defense of Moscow.

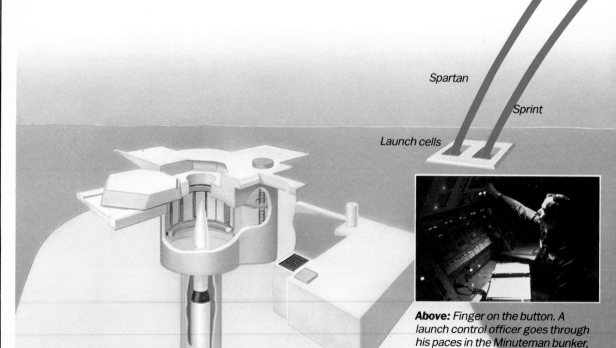

Spartan

Sprint

Launch cells

Above: Finger on the button. A launch control officer goes through his paces in the Minuteman bunker, Ellsworth Air Force Base, South Dakota.

Satellite knocked out by EMP

Target track radar

Missile track radar

Widespread destruction of electronic circuitry in city

Acquisition radar

Discrimination radar

Above: The unforeseen effects of EMP or Electromagnetic pulse were a major factor in the decision to scrap Project Safeguard. EMP is generated by transfer of energy from gamma rays to electrons when a nuclear weapon explodes. The explosions as Spartan and Sprint **(left)** intercepted incoming missiles would have disabled important communications systems and logic circuits all across the US. EMP also cripples satellites, whether they belong to friend or foe.

Anxiety about the Soviet FOBS program accelerated the development of the Thor anti-satellite. This Thor is ascending from Vandenberg Air Force Base, one of two sites (the other being Johnston Island) closely connected with the US effort to countermeasure Soviet orbital weapons.

called 'launch platform' during the early hours of the next day.[84]

Of the nine flights made during 1967, only one (Cosmos 160) appears to have failed. Royal Aircraft Establishment tracking data indicates that on this occasion the warhead broke into about 16 pieces during the first orbit, the debris re-entering some 19 hours later.[85] Whether this was the deliberate destruction of a malfunctioning payload or an accidental explosion has never come to light.

Such an unusual flight profile attracted Western attention. A number of theories were put forward. Some held that it was part of the orbital weapons program; others, that it was to test the effectiveness of missile defense radars and Soviet ICBM penetration aids (decoys meant to mislead Western ABM radars). One strongly held belief in some quarters was that the flights were connected with the testing of the Soyuz manned spacecraft recovery systems — the Soyuz 1 flight in April of the same year had crashed when the recovery parachute became tangled.[86]

Later there was speculation that the 'unusual' flights were testing a quick-reaction reconnaissance satellite.[87] There was uncertainty within the US Intelligence community. An orbital weapon was suspected but not until October was the evidence conclusive.[88]

The Soviets showed the FOBS SS-9 launch vehicle for the first time during the Red Square parade on November 7, 1967, describing it as capable of both intercontinental and orbital flights. On November 16, they responded to McNamara's statements on the subject by calling them a 'fabrication'.[89]

After the hectic pace of 1967, the FOBS program assumed a regular two-flight-per-year pattern, indicating that the system was considered operational. These launches were probably mainly for crew training and proficiency but they also served to provide information on the reliability of the system.

Beginning of the end for FOBS

The operational FOBS missiles were deployed in an isolated launch complex to the west of the Tyuratam test center. Eighteen FOBS silos were built, a small number in comparison with the SS-9s of which more than 300 were deployed.[90,91] Six months separated the last FOBS flight in 1967 and the first operational launch of Cosmos 218 on April 25, 1968. The year's second FOBS flight was Cosmos 244 on October 2. These two launches conformed to the typical FOBS profile of a single orbit.

Cosmos 298 on September 15, 1969 continued the normal pattern. The second flight, Cosmos 316, launched on December 23 showed characteristics similar to those of the un-

Space Law 1957–1967

The great technological strides of this decade meant that agreements banning space weapons were urgently needed. Beginning even before Sputnik, attempts to construct space law spanned three US administrations and as many changes of leadership in the USSR. The chart outlines the evolution of space law during this critical period.
It displays the technological triumphs and the quarrels of the superpowers inside and outside the forum of the UN, alongside other significant events.

1957

EISENHOWER

Eisenhower in a State of the Union address announces that
    'we are willing to enter any reliable agreement which would mutually control the outer space missile (ICBM) and satellite development'. (Jan 10)

BULGANIN

Soviets launch Sputnik 1 – the first artificial satellite placed in Earth orbit (Oct 4).

Sputnik 2, with dog Laika on board, is launched (Nov 3). Sputnik 3, a flying lab, follows.

1958

Moratorium on nuclear weapons tests comes into force.

EISENHOWER

Eisenhower writes to Soviet Premier Bulganin,
    'outer space should be used only for peaceful purposes. We face a decisive moment in history in relation to this matter. The time to stop is now'. (Jan 12)

After a series of failures, the US successfully places the Vanguard satellite in orbit. (Mar 17)

BULGANIN/KHRUSHCHEV

Soviet reply calls for ban on orbital weapons and
    'simultaneous elimination of foreign bases on the territories of other countries, first and foremost on the territories of the countries of Europe, the Middle East and North Africa.'

Terms unacceptable to the US. (Mar 15)

1959

EISENHOWER

The first US reconnaissance satellite, Discoverer 1, is launched. (Feb 28)

1960

Geneva Ten Nation Disarmament Conference stresses need to ban space weapons. (Mar 16)

Paris Summit Conference.

EISENHOWER

A U-2 reconnaissance aircraft flown by US pilot Gary Powers is shot down over Soviet territory. (May)

KHRUSHCHEV

Valerian Zorin, Soviet representative at the Disarmament Conference, flatly rejects a ban on orbital weapons unless the US agrees to give up its overseas bases. (April 4)

1961

EISENHOWER/KENNEDY

John F Kennedy addresses UN and proposes a ban on nuclear weapons in space or on celestial bodies. Vice President Johnson calls on UN to make
    'at least a brave start' toward banning space weapons.

KHRUSHCHEV

Major Uri Gagarin aboard Vostok 1 is the first man to orbit the Earth (April 12). Titov is next in Vostok 2. (Aug)

Series of nuclear-test explosions in Earth's atmosphere, including a 50-megaton test. (Oct 30)

1962

Soviets join the reorganized UN Committee at Geneva on the peaceful uses of outer space, presenting (on June 7) a nine-point declaration of principles as a guide to space activities. The proposals also feature a ban on space reconnaissance.

Terms unacceptable to the US.

KENNEDY

Thor shots made in which a nuclear warhead is detonated in space, generating radiation damaging to satellites which is trapped in the Earth's magnetic field. Starfish shot July 9, Blue Gill shot Oct 26, Kingfish Nov 1.

KHRUSHCHEV

Soviets fly their first reconnaissance satellite, Cosmos 4. (Apr 26)

Soviets detonate test bombs over Soviet central Asia, with side effects along lines of the US Thor shots. (Oct 22, 28 and Nov 1)

1963

Limited nuclear test ban treaty negotiated between the US and USSR.

Private US and USSR negotiations over guiding principles for space law. Agreement in principle is reached (Nov 7) and the text is released at the UN.

UN General Assembly passes a space weapons resolution calling on all nations not to place nuclear weapons in orbit. It is unanimously adopted by the General Assembly (on Dec 31). This is only a resolution, however, and not binding.

KENNEDY/JOHNSON

KHRUSHCHEV

Khrushchev, at a mass reception in Red Square for the Vostok cosmonauts, calls for space to 'become a zone of peace and a zone of international co-operation.' (June 22)

Soviet Foreign Minister Andrei Gromyko stresses the need in an opening address at the UN General Assembly for the US and USSR to reach agreement on orbital nuclear weapons. Significantly, his proposal does not insist on the removal of US bases from other countries. (Sept 19)

The Soviets drop their demand for a ban on satellite reconnaissance.

1965

US and USSR discuss at length drafts of a treaty based on these and the general resolutions of the UN. (Nov 1963 to Dec 1965)

JOHNSON

KHRUSHCHEV/BREZHNEV

1966

The UN General Assembly unanimously approves (on Dec 19) the Treaty on Principles Governing the Activities of States in the Exploration and the Use of Outer Space, including the Moon and other Celestial Bodies.

JOHNSON

Lyndon B Johnson announces that the US and USSR are in agreement regarding space weapons. A treaty is made
    'not to place in orbit around the Earth any objects carrying nuclear weapons or any other kinds of weapons of mass destruction.'

BREZHNEV

1967

Precisely four years after the limited test ban treaty, the space treaty comes into force. (Oct 10).

JOHNSON

Johnson describes the treaty as the 'first firm step towards keeping outer space forever free of the implements of war.'

BREZHNEV

At ceremonies in Washington, London and Moscow, the representatives of 62 nations sign the treaty. It is ratified (on May 19) by the Presidium of the Supreme Soviet.

SS-18

announced FOBS precursors of 1966, without the explosions. The second stage was left in a 982 by 91-mile (1581 by 147-km) orbit. The 'platform' went into a 572 by 81-mile (920 by 130-km) orbit and the payload finally assumed a 1025 by 96-mile (1650 by 154-km) orbit. Atmospheric decay slowly lowered this. After 248 days, on August 28, 1970 it re-entered.

As luck would have it, this occurred over the US midwest. The payload broke up into several pieces, landing in Oklahoma, Kansas and Texas. Some of the pieces were over 3 ft (1 m) in length and weighed in excess of 20 lb (10 kg). Certain sources maintain that pieces of the debris resembled a bomb casing, although this was never confirmed.[92]

The two FOBS shots of 1970, Cosmos 354 (July 28) and Cosmos 365 (September 25) reverted to type. The last FOBS launch was Cosmos 433 on August 8, 1971 and this flight ended the program. There had been 18 FOBS orbital flights in all — 2 precursors, 9 test flights, 6 operational launches and the mysterious Cosmos 316 mission.

Why was the program concluded? Had the Soviets come to agree with McNamara that FOBS was too inaccurate to be a practical weapon? That might explain the token number deployed. Or was the halt a political decision connected with the SALT 1 negotiations? It is true that when the SALT 1 treaty was signed, it limited the US and USSR to two ABM sites each — one to protect ICBM sites, the other the national capital. This was later cut to one site each, with a total of 100 launchers. On October 2, 1975 the US Congress directed that the single US ABM site be closed down, removing as a result the prime FOBS target.[93]

The SALT II treaty banned the development, testing and deployment of Earth-orbital nuclear weapons. FOBS was specifically included in this ban, and of the 18 FOBS launchers at Tyuratam, 12 were to be destroyed. (The other six were available for conversion into test pads for a new generation of conventional missiles.) Destruction of the 12 was a condition, and this had to be done within eight months of the treaty's ratification. Conversion of the remaining six could take place after the treaty came into force.[94] Although the SALT II treaty was never ratified by the US, both countries agreed to follow it. Between 1975 and 1980, the Soviets retired the entire SS-9 force, replacing it with SS-18 ICBMs.[95]

Mercifully, with this act orbital nuclear weapons passed into history. They were abandoned on grounds of their expense, inaccuracy and unreliability. Economics governed this weapons race as every other and the fact that Khrushchev's military objectives could be pursued only at the expense of his domestic ones was crucial. Initiatives to limit and even ban military activity in space were made even before Sputnik I, but while technology advanced at a pace unprecedented in human history, effective space law was evolving very slowly.

Chapter 3

THE ORIGINS OF ANTI-SATELLITES

'(…) to insure that no nation will be tempted to use the reaches of space as a platform for weapons of mass destruction, we have (…) developed (…) systems to intercept and destroy armed satellites circling the Earth in space. I can tell you today that these systems are in place. They are operationally ready and they are on alert to protect this nation and the free world. Our only purpose still is peace (…)'

President Lyndon B Johnson, speaking at Sacramento, California on September 17, 1964

The Soviet Union, then, had demonstrated in the early 1960s that it had the technology to construct orbital nuclear weapons. The Vostok missions, the superbomb tests and Soviet successes with large boosters all gave Khrushchev's threat at the Titov reception a distinctly plausible tone. An anti-satellite (or ASAT) system was now imperative.[1] Its role was twofold. First, it was argued that a countermeasure such as this was needed to enforce any agreement banning space bombs.[2] Second, it would provide a means to destroy a hostile nuclear weapon in orbit, before such a weapon could strike its intended target.

Since the system was required within a year or two years at the outside, there was no time for the sophisticated approach of orbital rendezvous. Attention therefore shifted to a direct-ascent attack profile. The plan was to launch an ASAT from the ground or from an aircraft; it would then follow a ballistic flight path to close in on and demolish its target. This approach had the advantage that it utilized existing technology, whereas orbital rendezvous (the placing of an ASAT in orbit to 'kill' an enemy satellite) would call for a pioneering development effort. The latter might cost $1 billion as against the estimated cost of only $100 million for a ballistic ASAT.[3]

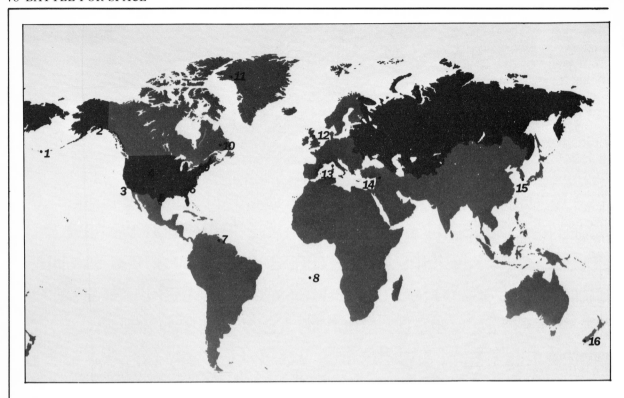

NORAD's Space Detection and
Tracking System. Soviet missiles
register first on the BMEWS radars
in Alaska, Greenland and England.
This data is supplemented by radars
in Turkey and the Aleutians in the
North Pacific. The Air Force tracking
network is used alongside SPASUR
to calculate satellite orbits. An
effective ASAT system demanded
several improvements to this web of
radar and optical sensors.

1 Shemya
2 Clear
3 Edwards
4 Cold Lake
5 NAVSPASUR
6 Elgin
7 Antigua
8 Ascension
9 Milstone
10 St Margarets
11 Thule
12 Fylingdales
13 San Vito
14 Diyarbakir
15 Pulmosan
16 Mt John

Ground-based radars and telescopes were now doubly important for target identification and tracking. Improvements in the US tracking network operated by the North American Air Defense Command (NORAD), a system known as SPADATS (Space Detection and Tracking System), were called for.

Soviet launches were picked up first by the Ballistic Missile Early Warning Radars at Clear in Alaska, Thule in Greenland and Fylingdales Moor in England. Radars at Diyarbakir, Turkey also supplied data and a radar station on Shemya Island in the Aleutians tracked Soviet satellites as they went into orbit. These findings were collated with those of the Air Force tracking network and the Navy's Space Surveillance System (SPASUR) in order to calculate the orbit in question. A series of transmitters and receiving stations stretching in a line from Fort Stewart, Georgia to San Diego, California were also involved. These transmitters send out vertical fan-shaped beams, forming an electronic fence which spans the southern United States. Whenever a satellite crosses the fence, its echo is detected by the receivers.[4,5]

To support a direct-ascent ASAT new radars would have to be added to the SPADATS network, accuracy improved and data-taking and processing techniques refined. Procedures to relay target data swiftly to the ASAT launch site were another important priority.[6]

The role of radar in making an ASAT strike possible can hardly be overstressed. A weapon-carrying satellite must be identified by the characteristics of its orbit, decay rate, analysis of its telemetry and its orbital behavior. Departures from earlier flight patterns and the use of new orbits and inclinations are tell-tale signs. Known as radar signal analysis, the technique of bouncing signals off an orbiting satellite provides an indication of a satellite's size and shape. The radar return from a cylinder, for instance, will be different from that of a flat panel.

Processing this data is difficult since the radar return from a cylindrical part of a satellite must be separated from that of a cone-shaped section and again from the oblong solar panels. Each element has to be isolated and identified. The size of an object (or its radar cross section) is determined from the strength of the returning echo. If the signals show large and rapid changes, the object is tumbling. A satellite's material composition can even be determined from returning radar signals. Some materials absorb radar signals and so reflect poorly, whereas aluminum or steel reflect well.[7]

Tracking satellites by telescope

Ground-based telescopes are also used for space tracking. Funding for such a telescope was given by the Air Force in 1960 and the design was completed in 1961, producing an F 6.5 Newtonian telescope with a 48-in (122-cm) diameter main mirror. To track accurately a swiftly moving satellite against the star background the telescope has an unusual mounting which allows rotation as well as altitude and azimuth adjustment.[8]

When work began on the telescope, the intention was to install it near the Wright-Patterson Air Force Base but city lights, smog and aircraft vapor trails forced a location change. The site finally selected lies near Cloudcroft, New Mexico amid the pine trees of the Lincoln National Forest. It is at an elevation of over 9000 ft (2743 m) and can take advantage of dark skies and excellent viewing conditions.[9]

With its large dome, the building resembles an astronomical observatory. At opposite corners of the roof are two smaller domes which house 5-in (12.7-cm) spotting telescopes. An observer seeks the target satellite in the first instance through one of these spotting telescopes. As the main telescope is electronically tied to the spotting telescope, its movements follow those of the smaller one. Once the main telescope has acquired the target, it tracks the satellite automatically. The observer records his visual impressions while photos are taken with a 35-mm camera and special high-speed film. The first tracking of a satellite by the Cloudcroft telescope took place in February 1964.

The average resolution of the telescope is 1 arc second. This is equivalent to 3.8 ft at a 150-mile altitude (1.16 m at 241 km), enough for an observer to see a wealth of detail. On a

Saturn 1 upper stage observed in this fashion it is possible to discern quite clearly discoloration from the rocket's firing. By way of a test, a converted fuel tank launched by a Titan IIIC was painted with black lines to try Cloudcroft's resolution. Over the course of two weeks, an observer if he wished might see them gradually fade as ultraviolet radiation eroded the paint.

In addition to photographic observation, the Cloudcroft telescope was equipped to conduct infrared scans of a satellite and make laser-ranging distance measurements. To observe satellites unlit by the Sun, the telescope had a low-light television camera and a laser-illumination system. Satellites were ordinarily photographed during the early evening or before dawn while still illuminated by the Sun below the horizon: the Cloudcroft telescope could observe satellites throughout the night. The telescope was not operated by NORAD but by the Air Force avionics laboratory.[10] The information it provided would assist in determining the function of a satellite.

A direct-ascent ASAT remained a complex technical task. The speeds involved were such that if the rocket were to reach the interception point even a fraction of a second too early or too late, the miss distance would be several miles. For this reason, a nuclear weapon was usually required as the 'kill mechanism'.

Nuclear explosions in space

A nuclear detonation in space differs significantly from one in the Earth's atmosphere; the most important difference being that there are no shock waves in the vacuum of space. However, three characteristics of a nuclear explosion make it effective in an anti-satellite role. These are the thermal flash from the fireball, hard radiation, and the electromagnetic pulse (or EMP). A nuclear explosion generates a brief but very intense flash of light and heat, weakening or burning through a satellite's structure even at a distance, and/or overheating electronic components and disabling infrared horizon sensors and temperature control systems. The subsequent fireball emits radiation primarily in the form of X-rays and neutrons.

When this radiation penetrates a transistor it creates ions which accumulate and alter the transistor's electrical properties disastrously. Guidance and arming circuits could be disabled in this way and solar cells are similarly at risk. Radiation absorbed by the ablative material on the re-entry vehicle may generate enough heat to actually melt the heat shield.[11] The flash and radiation from such an explosion are enhanced in space because there is no absorption by the atmosphere.

The last and most subtle effect is the electromagnetic pulse. The hot ionized gases of the fireball, cutting the Earth's magnetic field, produce long-standing waves. These waves induce extremely high currents in such electrical components as straight wires or cables. Solid-state components such as transistors, diodes and integrated circuits have very narrow tolerances to stray currents. An electromagnetic pulse can even cause current to flow in materials which are ordinarily electrical insulators, and the surge of current which it induces plays havoc with solid-state components.[12,13]

When the US Department of Defense (or DoD) started to study ASATs in the early 1960s, there were several possibilities in the shape of boosters. One was the Skybolt air-launched ballistic missile, a two-stage solid-fuel rocket with a horizontal range of 1150 miles (1850 km) which could be launched from a B-52. Skybolt was quickly eliminated (in December 1962) due to a string of failures, changes in US nuclear policy and the missile's failure to meet cost-effectiveness standards.[14,15]

Another candidate for the ASAT booster was the Minuteman ICBM which used solid fuel and so avoided a long fueling and countdown procedure. It was at this time only in the development stage, however, and so was passed over. (The first Minuteman ICBMs were deployed operationally in late 1962.)

Th⸱ ... titled Early Spring ... om the submarine ... several respects. ... uclear warhead. T ... submarine, the boo ... o attain the altitude ... e predicted height ... e came into play. U ... vehicle was intended to boost for ... seconds, ... r the target satellit ...

T⸱ ... a guidance unit w⸱ ... satellite and its bac⸱ground ... ned a wide field in ... ise tracking and gu⸱ ... rget that the unaide⸱ ... d be relayed to the ⸱ ... n. After the Early ... he warhead would ... nd finally a proxim⸱ ... lets.[18] The impact ... pellet and a satellit⸱ ... d complete destru⸱ ...

S⸱ ... t to insure a high '⸱ ... submarines were ⸱ ... ite's ground track ⸱ ... h sites.[19] In addition to studies, the optical tracker of the guidance system was tested. A sound technical notion ahead of its time, Early Spring fell foul of costs like many such schemes.[20]

In the end, there were two alternatives: the Army's Nike-Zeus Anti-Ballistic Missile; and the Air Force Thor Intermediate Range Ballistic Missile. Both were developed, though only one saw long-term service.

The life and hard times of Program 505

Program 505, the first operational ASAT, had a brief and ill-starred life. It was bedevilled by its rival, Program 437, from the start. Program 505 was the adaption of the Army's Nike-Zeus Anti-Ballistic Missile, while Program 437 was an ASAT using the extremely versatile Thor rocket.

The technology of an ABM is readily adaptable for an ASAT role since both types of missile must be able to intercept targets in space. There are, however, differences in the geometry of the interceptions; an ICBM warhead follows a descending trajectory as compared with a more horizontal satellite path. Differences in angle, distance and speed of the target must be allowed for when programming the guidance radar, but these differences are relatively minor.

An ABM must likewise be ready to fire at a moment's notice. Indeed, in some ways an ASAT's role is less demanding than that of an ABM since only one target would be engaged at a time. From orbital predictions, the target's flight path and direction, time of appearance and altitude would be known precisely many hours in advance. An ABM on the other hand must cope with multiple targets, decoys, jammers and booster debris simultaneously, with the briefest possible warning.

Work on the Program 505 Nike-Zeus began in February 1955, the first test launch being made at White Sands Missile Range on December 16, 1959. Nike-Zeus stood an impressive 50 ft 2 in (15.29 m) tall and had three solid-fuel stages: the booster or first stage; the sustainer or second stage; and the nose (also called the jet head). The nose contained the warhead, guidance unit, hydraulic pump and control nozzles. Its thermonuclear warhead had a yield under 1 megaton.[22,23]

The Nike-Zeus was guided to its target by two separate radars and a ground computer. The Target Track Radar picked up the ICBM warheads and determined their flight path using a precise and narrow radar beam to follow these small, high-speed targets. The trajectory of the Nike-Zeus as it climbed was monitored in turn by the Missile Track Radar. (A beacon was carried aboard the Nike-Zeus for just this purpose.) Data from these radars was fed to the Target Intercept Computer which used the tracking information to calculate missile steering commands transmitted through the Missile Track Radar.

The ASAT potential of the Nike-Zeus was recognized early. In March 1961, Brig General David C Lewis, Army Director of Special Weapons Development claimed that with minor modifications the Nike-Zeus could destroy satellites at an altitude of 200 miles (321 km). With major modifications, it could reach 1200 miles (1930 km). This extended-range version could be ready by 1967 with normal development and funding[21]. General Lewis' timing was well-chosen, for defense officials were becoming very concerned about the pace of Soviet space activities.

By the time Nike-Zeus left the atmosphere it would therefore have the correct trajectory to intercept its satellite target. After the second stage cutoff, the jet head would separate and coast for a few moments. Seconds before the interception, the Target Intercept Computer ignited the jet head's motor and sent it homing towards the target. The warhead then exploded. Nike-Zeus was the first missile equipped to make a 'satellite kill' in this way.[24]

In early 1962, McNamara took an interest in the Nike-Zeus ASAT. Bell Telephone Laboratories — one of the Nike-Zeus contractors — reported that it could be ready in about a year. McNamara approved the program in May 1962.[25]

The Nike-Zeus ASAT was based at Kwajalein Atoll in the Pacific, a Nike-Zeus facility having been built there for ABM test flights against US ICBMs launched from Vandenberg AFB. The facility included complete test and launching equipment and a full set of radars. DM-15B series missiles were modified to meet the higher altitude and longer range demands of an ASAT mission. These modifications included Polybutadiene Acrylic Acid solid-fuel propellent which would give better booster performance, battery capacity was increased from two to five minutes to cover the protracted flight time and a new higher-capacity hydraulic pumping unit was added. This new version was designated DM-15S.

Development was fast. The first Nike-Zeus ASAT was ready seven months after the formal go-ahead. The first DM-15S was fired from White Sands Missile Range and successfully intercepted an imaginary target in space at an altitude of 100 nautical miles (185 km), bringing it within the lethal distance of the nuclear warhead. (This missile incidentally set an altitude record for a White Sands-launched Nike-Zeus.)

The first US ASAT nears completion

In the meantime, the Air Force had started work on a Thor ASAT and various developments had been proposed. The Army was to complete work on Program 505 and be responsible for any Nike-Zeus successors while the Air Force was to build a high-altitude Thor ASAT and conduct research into advanced systems, including a non-nuclear ASAT and satellite inspectors. The Navy intended to look into possible space threats to its own activities.[26]

A second Nike-Zeus ASAT was fired from White Sands on February 15, 1963, intercepting an imaginary target in space at a 151-nautical mile (279-km) altitude. Tests then shifted to Kwajalein Atoll, from where the third DM-15S missile was fired on March 21, 1963. Its target was this time a simulated satellite in a 112-nautical mile (207-km) orbit. However, on this occasion, the Missile Track Radar failed to lock on and the mission was unsuccessful.

A British RAF Thor lifts off from Vandenberg AFB. The Thor IRBM has an interesting history. Work began on it in November 1955, when such a missile was needed to stand in until the Atlas and Titan ICBMs (then in development) were ready. It was 65 ft (19.8 m) long and 8 ft (2.44 m) in diameter, weighing 110 000 lb (49 869 kg) at launch, and had a single Rocketdyne LR 79 engine providing 150 000 lb (68 040 kg) of thrust. In 1959, Thor missile squadrons were grouped around four RAF bases in Britain including Feltwell in Norfolk – their range of 1976 miles (3180 km) including the western USSR up to an arc drawn from Murmansk to the Crimea. These missiles were armed with H-bomb or thermonuclear warheads.[34,35]

Kwajalein Atoll in the Pacific is the main US ABM test site and home of the Nike-Zeus Program 505 ASAT. The launch site is in an artificial hill at the far end of the island.
1 Launch cells
2 Missile track radar
3 Target track radar
4 Discrimination radar
5 Airstrip
6 Acquisition radar receiver

Things also went badly on the next flight. A launch on April 19 failed when the beacon carried by the Nike-Zeus went dead 30 seconds before the interception and so miss distance from the simulated satellite in its 160-nautical mile (296-km) orbit could not be determined.

In the next test an Air Force 162A series Agena D was equipped with miss distance-measuring instruments and systems to improve its radar return.

The Target Track Radar picked up the Agena D at long range and began precise tracking. A specially fitted-out Nike-Zeus was then fired in pursuit; stage separation was normal, the jet head and its simulated nuclear cargo climbing toward the Agena D. When the Target Intercept Computer ordered detonation, it was well inside the weapon's lethal radius and had the exercise been authentic the satellite would certainly have been destroyed.[27]

McNamara met with his senior advisers to set policy for the US ASAT program on June 27, 1963. Decisions taken at this meeting affected Program 505 in two ways. McNamara directed the Army to complete the missile installation (including warhead supplies) and have it ready for fast-reaction interception. At the same time, he argued that Program 505 was in direct competition with the Air Force's high-altitude Thor ASAT and that both could not be maintained. The Air Force and Thor ASAT system were favoured and Program 437 was deemed a higher priority than 505.[28]

Several DM-15S missiles were stockpiled on Kwajalein Atoll, one always being at operational status and ready to fire. The Army had a nuclear warhead on permanent standby. Program 505 acted as a fast-reaction ASAT able to neutralize satellites in low orbit, and the program was staffed by Army personnel. Douglas and Bell Laboratory personnel strove hard to trim even further the time needed to prepare the weapon for launch, with Program 505 launches taking place throughout 1964.[29] In one of these, a Nike-Zeus was fired against a simulated satellite. The range was 93 nautical miles (172 km) from Kwajalein Atoll at an altitude of 79 nautical miles (146 km). Miss distance was determined to be 708 ft (216 m). In a real interception, the target would have been enveloped by the expanding fireball.[30]

McNamara finally directed in May 1966 that Program 505 be phased out. Whatever advantages the Nike-Zeus may have had in reaction time, the superior altitude of Program 437's Thor made it redundant.[31]

Program 437 — the way forward The main objective of the US Air Force direct-ascent ASAT program which began on February 9, 1962 was to acquire the ability to destroy satellites during a cold war. The booster concerned had to employ either solid fuel or storable liquid propellent, stressing system-effectiveness and simplicity. The operational ASAT system must be economical to support and maintain, and able to intercept satellites to the limit of the tracking system while using a minimum number of launch sites. Reaction time had to be as short as operationally reasonable and both nuclear and non-nuclear warheads considered. War-time conditions were to be contemplated and attention given to protecting the tracking system. The ASAT should also be able to undertake covert attacks as well, but these missions should not pose any danger to the public from debris or weapon effect. The Air Force Systems Command was ordered to investigate the alternatives.[32]

Meanwhile, the Air Defense Command (or ADC) was quite separately considering ASATs. The result was an ambitious three-phase program designed to deliver a prototype operational ASAT by 1964.[33]

Air Force Systems Command had completed its studies in autumn 1964, delivering a proposal centered on the Thor IRBM fitted with a nuclear warhead as the 'kill mechanism'.

The launch site for this Thor ASAT was Johnston Island, a remote spot in the Pacific Ocean 715 miles (1150 km) south of Honolulu. Only a mile long and a quarter-mile wide (1.6 by 0.4 km), it normally served as a refuelling stop for Air Force transport aircraft. A Thor facility was built there in 1962 for high-altitude nuclear tests as part of the Dominic series.[36] (On July 9, 1962 the Starfish shot took place and a Thor-launched 1.4-megaton warhead was detonated at an altitude of 250 miles (402 km). Further tests at lower altitudes were made later.[37,38])

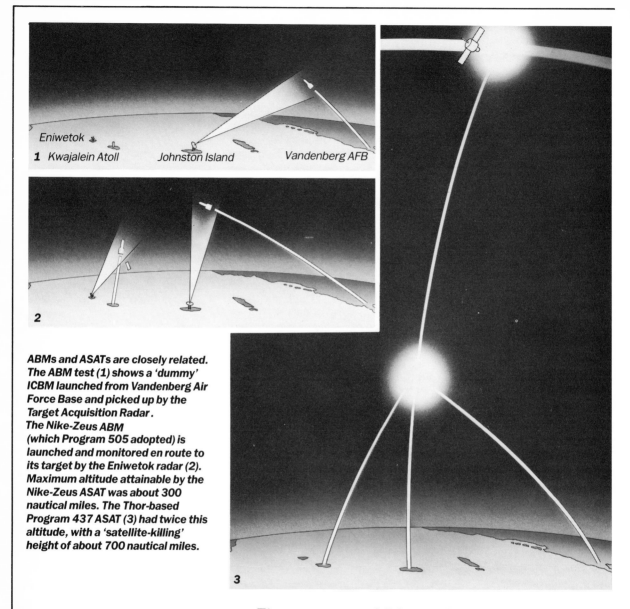

ABMs and ASATs are closely related. The ABM test (1) shows a 'dummy' ICBM launched from Vandenberg Air Force Base and picked up by the Target Acquisition Radar. The Nike-Zeus ABM (which Program 505 adopted) is launched and monitored en route to its target by the Eniwetok radar (2). Maximum altitude attainable by the Nike-Zeus ASAT was about 300 nautical miles. The Thor-based Program 437 ASAT (3) had twice this altitude, with a 'satellite-killing' height of about 700 nautical miles.

The remoteness of Johnston Island ensured safety and security. The biggest problem was that the Thor's propellents were liquid oxygen and RP-1 kerosene and therefore the missile would have to be fueled during the terminal count-down. Once fueled, it could not be held indefinitely. This was a drawback, but the Thor was the only missile at hand with the necessary altitude.

General Bernard Schriever of Air Force Systems Command was enthusiastic about the program and he called on ADC to render the system operational. Defense Secretary McNamara was briefed on the Thor ASAT in November, 1962. In December the Secretary of the Air Force was informed that:

'some version of your proposal (the Thor ASAT) appears

to be the fastest way of obtaining an increased capability on range and altitude beyond that which will be available from the Nike-Zeus installation on Kwajalein (Program 505).'

The Thor ASAT program development plan was limited to a four-launch feasibility demonstration scheduled for early 1964. It received $11 million in the fiscal 1964 budget, $6 million being reprogrammed from the budget of fiscal year 1963.

Program 437 became the Thor ASAT's official title and its scope was expanded; the Air Force Systems Command was directed to render it operational in the event of an emergency and ADC personnel were assigned to all levels of the development effort.[39] The goal for reaction time, from alert to launch, was two weeks. The attempt was to improve this with existing personnel and resources.[40]

Program 437's working-plan was published on March 14, 1963. A minimum number of people would be permanently stationed on Johnston Island, maintaining the two-pad launch complex. Most of the personnel, missiles and other equipment would be at Vandenberg AFB, three missiles (two primary and one spare) being kept at operational status ready for emergency air transport to Johnston. The nuclear warheads, spare parts and tools were packed into transportable kits.

On detection of a hostile satellite, the unit would be alerted, national authorities would be notified and missiles readied. Two missiles and their kits would then be airlifted to Johnston Island. The crew on Johnston Island would simultaneously begin pre-launch preparations. Both missiles would be counted down, one of them acting as the primary interception vehicle. In the event of the primary missile suffering a launch failure, the second missile would be fired toward another orbital point to intercept the hostile satellite.

Program 437 won new support two weeks after the working-plan's publication. On March 20, 1963 Secretary of the Air Force, Eugene M Zuckert, told General Curtis LeMay the Air Force Chief of Staff, that an operational ASAT capability had top priority among defense programs. General LeMay was instructed to supply the Air Force Systems Command with everything necessary for research and establishment of an emergency ASAT system.[41] Personnel with Thor experience were assigned to the project. Brockway McMillan, Assistant Secretary for Research and Development recommended that the reaction time of Program 437 should be limited only by the technical features of the system, not by cost. A two or three-day reaction time was aimed for.[42]

Program 437 was gradually metamorphosing from an emergency standby system into an operational ASAT on permanent alert. This became official at a meeting called by McNamara on June 27, 1963. He criticized the planned reaction time of Program 437 and made it clear that he wanted

Above: Resembling an aircraft carrier, Johnston Island basks under a tropical sun. Its remoteness from other land made it ideal as a base for the Program 437 ASAT.
Right: The Program 437 area on Johnston is just above the square black parking area. The two pads are visible. Also apparent is the launch control center (the low mound near the beach) and the two Ground Guidance Stations (the tall square buildings). The center of the island is the housing area for the combat crews. In the foreground is the dock where several landing craft can be seen. The island's harsh climate took its toll on Program 437, and much of the damage caused by Hurricane Celeste was from humidity corrosion.

the ability to initiate destruction of a satellite with a phone call.

A serious threat to Program 437 appeared in the spring of 1963, however. It had not been spelled out amid all the changes just who would operate Program 437 once the test flights were over: Air Force Systems Command or Air Defense Command? This uncertainty caused increasing stress as the Program expanded, threatening to make it a casualty of bureaucratic infighting. ADC Commander General Robert Lee requested the Air Force Chief of Staff to name ADC as the using command to remedy this and his urgent request was answered.[43]

The Air Defense Command and Air Force Systems Command were next ordered to develop a plan together for the orderly transfer of responsibilities. All formal training was completed by November 15, 1963 on which date the 6595th Test Squadron was transferred from Systems Command to ADC, deactivated and reconstituted as the 10th Aerospace Defense Squadron expressly to man Program 437. The unit had an authorized strength of 49 officers, 451 airmen and 3 civilians.[44]

The satellite killer in action

Program 437 demonstrates very clearly the umbilical relationship between politics and military spending. This pioneering anti-satellite effort spanned 13 years and was pursued intermittently, with frequent changes of status. In the presidential election year of 1964, Senator Barry Goldwater accused the US Administration of ignoring the threat from space weapons. His allegations prompted President Johnson to make the first public reference to the Nike-Zeus and Thor anti-satellites which were then in operation. Subsequent doubts in political circles about the usefulness of these systems after all and the siphoning-away of funds for the Vietnam war made the early US ASAT program a haphazard affair.

The blueprint for Program 437 was ready in August 1963. Thor rockets were kept on pads on Johnston Island, ready for launch. Two more waited in reserve at Vandenberg AFB. As before, there would be a dual countdown of the primary and secondary missiles. Program 437 centered around the LV-2D Thor which had an altitude capability of 700 nautical miles (1296 km) and a horizontal range of 1500 nautical miles (2778 km). The nuclear warhead — an MK 49 — had a 'satellite-killing' radius of 5 miles (8 km). Damage assessment of the target would be calculated by taking into account booster performance, miss distance, missile telemetry and tracking data.[45,46]

Anxieties surrounded the first flight. Although the Thor had proven itself previously, the launch crew and manage-

1 Launch control center
2 Ground Guidance Station 1
3 Ground Guidance Station 2
4 Launch Emplacement 1
5 Launch Emplacement 2

ment personnel could gain no comfort from the Thor flights during the 1962 Dominic nuclear test series when several missiles failed at launch. Early in the program doubts had been expressed, in the light of the Dominic failures, whether a four-shot test series would be sufficient.

The first Program 437 launch, conducted by contractor crews, took place on February 15, 1964[47,48]. The reserve missile was counted down alongside the primary, just in case. As the Thor climbed, telemetry and ground guidance data indicated a normal flight path to the intercept point. This time the target was real — a Transit 2A rocket body, occupying a 564 by 336-nautical mile (1045 by 622-km) orbit, inclined 66.7°. The intercept point was at an altitude of 540 nautical miles (1000 km) and a range of 443 nautical miles (820 km) from Johnston Island on an azimuth of 290°. The simulated warhead passed close enough to the target to be considered a successful interception.[49] The second test flight made on March 2, 1964 was also successful, as was the third on April 22, 1964.[50]

The last research and development flight, which occurred in May 1964, was conducted by personnel of the 10th Aerospace Defense Squadron. General Thatcher of ADC and General Funk of the Space Systems Division were on hand to observe the first launch by an Air Force crew, but unfortunately the Thor malfunctioned shortly after lift-off and went out of control. The trouble was traced to the booster rather than the Defense Squadron's launch procedures. Next day (May 29, 1964), the Squadron was declared ready. One missile and its nuclear warhead were constantly on hand. The second Thor and its warhead were flown to the island soon afterwards and on June 10, 1964 the Johnston Island facility was fully operational; both missiles were on 24-hour alert.[51]

Although there had been reports in the technical press that work was underway on a Thor ASAT, the scope of the program, the fact that flights had actually been made and that the system was operational remained secret.

This year witnessed Johnson's speech in California, when he made public US progress with ASATs. The Republican challenger in this election year was Senator Barry Goldwater and he had accused the Administration of ignoring military space systems and weapons. By disclosing the existence of the Nike-Zeus and Thor ASATs, Johnson refuted this potentially damaging charge. It was now common knowledge that the US could destroy enemy satellites of whatever description[52].

Maintaining the weapons system

Even after it is operational, a weapons system must undergo tests and exercises throughout the life of the system. In this way cumulative problems can be discovered and corrected. The first such test was made in November 1964 and was deemed a successful interception.[53] System Readiness Exercises of various kinds provided further information about Program 437's effectiveness. These ranged from a full-scale countdown of both fueled missiles, involving the NORAD Combat Operation Center and other elements of the complete system and ending with a simulated launch, to straightforward technical rehearsals. In practice sessions held independently on Johnston Island and at Vandenberg combat crews received training to keep their proficiency at a high level.

The second of the Combat Evaluation Launch (or CEL) series, carried out on April 5, 1965 also served to support the ASAT system's development. This time the target was the defunct Transit 2A satellite (placed in orbit in June 1960 and operational until August 1962). The interception was positive, preliminary data indicating that the miss distance of the simulated warhead was 0.89 nautical miles (1.65 km), a highly successful result.

Only two missiles for evaluation and crew training remained to cover five years. Without regular launches, the morale and skills of the combat crew were sure to decline.[54]

Air Defense Command therefore pressed for Thors. The DoD subsequently granted 18 additional Thors from mid-1965 through 1971, though half of these were for an Army program.[55]

To allow launches over a complete 360° circle around Johnston Island, Air Defense Command began to consider replacing the original Ground Guidance Station (a Bell Telephone Laboratory radar and an Athena computer) with a more sophisticated Titan 1 radar and Univac 642B computer.

A similar station would be built at Vandenberg Air Force Base for training and support. The Air Force's request for $188 000 in emergency funds to construct the station on Johnston Island was approved on March 8, 1966. Personnel began training at the Univac factory in July and plans were in fact drawn up for the prospective Ground Guidance Station at Vandenberg AFB. Work stopped abruptly on the Vandenberg GGS in early July 1966 when a halt was called to funding for the latter station while the Bureau of the Budget investigated the need. The original scheme for joint occupancy of GGS-6 was dropped thereafter. The new plan was to give ADC a separate training facility at Vandenberg AFB which would not be operational until late 1968[56]. In any event the new GGS on Johnston Island was completed by the end of 1966. Testing however revealed a radio interference problem between it and the launch pads.

Air Defense Command agreed to make use of the Ground Guidance Station at such time as its serious radio interference problem was solved. The delay caused by the interference problem had a useful consequence. It gave the combat crew time for the extra training which was felt necessary. SQUANTO TERROR 67-14 was designed to test the new Ground Guidance Station at the same time[57,58]. A piece of US orbital debris simulated a satellite armed with a nuclear weapon. Fired on March 30, 1967 the target was within lethal range of SQUANTO TERROR's warhead: miss distance was 2.019 nautical miles (3.74 km). The CEL proved that GGS-2 was operational. With this done, work could begin on converting the original GGS-1.

Each Ground Guidance Station was to have a Titan I AN/GRW-5 radar and Univac 642B computer. An informal ceremony took place at Vandenberg AFB to mark the transfer of Thor-Burner II launch operations to the 10th Aerospace Defense Squadron, a first step in easing the problems caused by the lack of CELs. The first 10th ADS Thor-Burner II ascended on June 29, 1967 carrying scientific satellites: the Army SECOR and the Navy AURORA 1. They were successfully placed into an approximately 2100-nautical mile (3889-km) circular orbit. Two more successful launches were made before the end of the year.[59,60] Most of these Thor-Burner II launches were of military weather satellites.

A revised schedule of practice launches was authorized

by the Secretary of Defense in December 1967. Funding would be provided for four training payloads in fiscal year 1968-69 and 70. Experience would be provided by the Thor Burner II and Program 922 test launches. (The latter as a non-nuclear ASAT planned to follow Program 437).[61]

December saw another setback in the shape of serious malfunctions within GGS-1's Univac 642B computer. To complicate things further, the problem of radar reflectivity was still unsolved.[62] The next exercise took place on May 15, 1968 by Combat Crew C and the launch was successful.[63] A second successful CEL was made later in the year on November 21.[64]

Although Congress had ratified the limited nuclear test ban treaty of 1963, the US decided to retain the ability to resume nuclear testing in the atmosphere in space and underwater if the USSR violated the treaty.

As a result, work began in 1968 on adapting the Program 437 Thor to allow nuclear tests in space as the High Altitude Program[65]. Another important decision taken at this time concerned the ASAT successor to Program 437.

It was a preliminary ASAT and the Air Force Defense Command envisioned a successor to correct its limitations. Foremost among these limitations was the use of a nuclear warhead, for reasons made plain by the Starfish nuclear test in space of 1962. The charged particles released were trapped by the Earth's magnetic field, creating an artificial radiation belt. Initial measurements indicated radiation 100 to 1000 times higher than before the test.[66]

Within a year the radiation levels had dropped to half their initial strength, though traces were expected to remain for ten years.[67] This trapped radiation was so strong that it damaged two Navy satellites (Transit 4-B and Traac), as well as the British satellite Ariel and Bell Telephone's Telstar 1.[68] If Program 437 was ever fired in anger, its radiation would cripple friendly and enemy satellites indiscriminately.

Decline of the early US anti-satellite systems
Various types of warhead were therefore to be investigated with a non-nuclear ASAT in mind.[69] By December 1963, the demands of Program 437 in terms of funds, facilities and manpower had grown and costs made a successor seem unlikely.[70] The Department of Defense fixed the ASAT budget at $3 million for fiscal year 1964 and $10 million for 1965, and attention shifted to adapting Program 437 for modest research and development efforts although the notion of a successor persisted in certain quarters[71].

The emphasis remained on a direct-ascent ASAT as opposed to an orbital rendezvous 'satellite killer', and boosters considered at this time included Thor and Titan III. The latter had the altitude to attack geosynchronous-orbit targets such as early warning and military communications satellites.[72] Enough progress was made in this direction for a successor to 437 to emerge in the form of the short-lived Program 922. Launched by a Thor, the interceptor was equipped with an infrared terminal guidance system and a high-explosive warhead.[73] Ling-Temco Vought was chosen in June 1967 to develop the system. Their letter contract was very preliminary, pending a determination of the amount of funding available. A sum of $20 million was allocated in fiscal year 1968, but in December 1967 half of this was transferred to meet Vietnam war costs and shortly afterwards the entire Program 922 was cancelled. Official doubts that a significant threat existed from space weapons — significant enough to justify the program's cost — were responsible.

Anti-ballistic missile defense was the revised aim of the Special Defense Program (SDP) which took up where 922 had left off[74]. Ling-Temco Vought remained the prime contractor and were told to concentrate on mid-course interception. Launch Emplacement 2 on Johnston Island would be modified for the SDP program. As well as trial runs of the interceptor vehicle, demonstrations would feature ground equipment, and tracking systems on Johnston Island in conjunction with NORAD Combat Operations Center.[75]

When SDP was first outlined in the spring of 1968, the first launching was planned for May 1969. Funding problems made this impossible. By the spring of 1969, the financial situation had improved and the first SDP launching was scheduled for December 16, 1969. In August, delays in obtaining components from subcontractors forced Ling-Temco Vought to request a postponement and accordingly the Air Force set a new launch date of March 1970. It was an all-too-familiar pattern.

Funding was not the only problem. The Air Force was actually running out of LV-2D Thors. After 1972, the supply was expected to be exhausted and older 'Long Tank' Thors would have to be used. Since these were taller than the others, time-consuming modifications would be necessary.

Air Defense Command was looking into a proposal that an emergency Program 437 launch capability be established at Vandenberg AFB, to insure against the Johnston Island facility being disabled in an accident. In the meantime, things had been happening at Johnston Island. The next CEL was scheduled expressly to test the new S-band telemetry system as well as overall performance. Program 437 telemetry had originally been transmitted on the L-band radio frequency (225-260 MHz) but the Air Force was eager to shift to a more secure frequency range to reduce the risk of enemy interference. Accordingly, all telemetry was to be converted to S-band (2200-2900 MHz) by January 1, 1970.

In the event, the CEL flight was not made until March 28, 1970 because of problems encountered by Vickers, the

S-Band contractor. The target was at an altitude of 580.4 nautical miles (1074.5 km) and a range of 902.7 nautical miles (1671.8 km): the total flight time was only 11 min 10 sec. The mission was considered successful.[76,77]

The first SDP launch took place on April 25, 1970 and failed disastrously. From the postmortem it appears that the Thor collided with the interceptor vehicle, damaging it in such a way that the cryogenic supply which cools the infrared sensors to increase their sensitivity to thermal emissions from the target was prematurely exhausted and the infrared guidance system failed to separate immediately. This expensive catastrophe brought the whole question of space and missile defense once more under scrutiny and the Special Defense Program quietly perished[78]. Six months later in an economy move Program 437 was placed on Standby status. Activities were limited to routine inspection and maintenance; 30 days would be needed to get it ready to make an interception.

The last nail in the coffin of the US direct-ascent antisatellite program was not a man-made disaster but a natural one. During the summer of 1972, in the lonely reaches of the Pacific, a storm began to develop which grew into Hurricane Celeste. Johnston Island was directly in its path. As the storm approached, the island personnel were evacuated to Hawaii. On August 19, the Hurricane passed about 21 miles northeast of the island, assailing it with destructive and sustained winds of 100 knots and gusts of as much as 130 knots (185 to 241 km/h). Huge waves, 30 to 40 ft (9 to 12 m) high, pounded the north and northeast sides of the island.

An air reconnaissance was made next day to assess damage in the storm's wake. Numerous expensive items had been exposed to water through damaged roofs or from condensation due to the loss of air-conditioning. All the computers were seriously damaged. After assistance from the Air Force and Univac had been requested, Univac representatives conducted a 'computer autopsy' on December 7, 1972 and Program 437 was removed from service the next day[79].

Why was Program 437 discontinued after this last setback? The odds against it had been steadily accumulating for some time. Very few of the LV-2D Thors which it required were available; its 30-day reaction time was too slow; and the nuclear warhead it favored was prohibited by the 1963 test ban treaty. (The Mk 49 warheads remained in storage until mid-1975, when they were returned to the Atomic Energy Commission for disposal[80].)

But perhaps the major failing of Program 437 in the view of the authorities concerned was that it did not and could not counter the threat posed by prolific Soviet military satellites. Consequently, the first chapter in the story of US ASATs was brought to a close on April 1, 1975 when Program 437 was officially terminated[81].

Chapter 4

THE ANTI-SATELLITE RACE

'Violence arms itself with the inventions of Art and Science in order to contend against violence. Self-imposed restrictions, almost imperceptible and hardly worth mentioning, termed usages of International Law, accompany it without essentially impairing its power.'

Carl von Clausewitz,
On War (1832)

Almost from the beginning of the space age, the Soviets found themselves facing a threat from space in the form of US photo-reconnaissance satellites. The US Air Force had begun low-level work on reconnaissance from space before Sputnik and in November 1957 the program was accelerated. The first US reconnaissance satellite was ready just over a year later. Discoverer 1 was launched on February 28, 1959 by a Thor-Agena. It suffered a malfunction and was the first of 12 consecutive failures. Success came at last with the Discoverer 13 launched on August 10, 1960 and a year later satellites were returning their film capsules on a fairly regular basis.[1] This of course represented a threat to Soviet security. They had worked long and hard to destroy the U-2, finally driving it from their skies, but reconnaissance satellites now laid bare the whole of the USSR.

Threats were made against the offenders. In June 1960, a month after the U-2 was shot down, Khrushchev warned that reconnaissance satellites could be destroyed in a similar fashion. Taking up the theme a few months later, the Soviet journal *International Affairs* condemned US spy satellites and echoed Khrushchev's remarks, asserting that 'the Soviet Union has everything necessary to paralyze US military espionage in outer space'.[2] Even the very low resolution photos from weather satellites were attacked as espionage.[3]

The Kennedy Administration clamped a news blackout on US reconnaissance satellites in November 1961 and the US Government would not even officially confirm that satellites were being used for reconnaissance.[4] The satellites' orbital characteristics were withheld for weeks or months after launch with the result that flights may well have ended by the time the information was published.[5]

Both of the Soviet UN proposals on space law featured bans on reconnaissance satellites and the Soviets frequently spoke out in the UN and other forums against US military space activities. At one point, they declared that the use of satellites 'for the collection of intelligence information (. . .) is incompatible with the objectives of mankind in the conquest of outer space'. The Soviets dropped their demand for a reconnaissance ban in September 1963, because they now had their own operational reconnaissance satellites — nine of them, to be exact. The first, Cosmos 4, went up on April 26, 1962.

They also attempted to build a legal case for the destruction of satellites. A book on space law published in 1962 claimed that such a right was 'indisputable', and supported it by reference to the doctrine of national sovereignty. Reconnaisance satellites supposedly violated a nation's sovereignty and in doing so posed a military threat. Under international law, a threatened nation had the right to defend itself.

The imagined threat of US orbital nuclear weapons also played a part in early Soviet ASAT planning, although this threat was less immediate than that posed by US reconnaissance satellites. Soviet spokesmen talked of the need to 'repulse an enemy attack through space'. The first edition of the Soviet book *Military Strategy* published in the summer of 1962 included a highly imaginative account of US space plans, exhorting its readers not to 'allow the imperialist camp to achieve superiority in this field'. The second edition (published in August 1963) stated that 'it is necessary to have corresponding means assuring the timely detection of enemy space equipment and its rapid destruction or neutralization'. In both instances a space weapon threat from the US was anticipated but no reference was made to an extant Soviet anti-satellite system. The 1963 edition referred to ASAT missions as a 'new problem on the agenda', and observed that 'it is still early to predict what line will be taken to the solution of this new problem (. . .)'[6]

The USSR traditionally places great emphasis on defensive measures. After World War II, the USSR was faced with the possibility of a massive nuclear air attack by the US Strategic Air Command and Britain's RAF Bomber Command. In response the Soviets established — as an independent branch of their armed forces — the National Air Defense Command (PVO-Strany).[7] Today it is a truly massive force with some 5000 early warning and height-finding radars, 12 000 SAM launchers and 2500 interceptor aircraft.[8] (US air defenses are minimal by comparison.) This defensive state of mind might well give rise to an anti-satellite program as soon as the threat

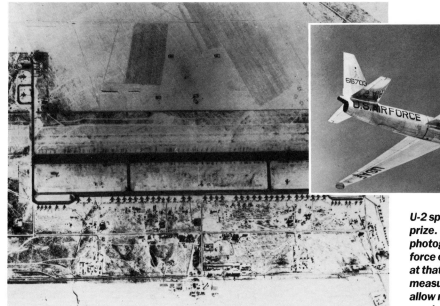

U-2 spyplane (inset) and its valuable prize. Taken in 1956, this CIA photograph shows nearly the entire force of M-4 bombers constructed at that time. Length and wingspan measurements calculated from it allow military photo-interpreters to reach conclusions about performance, range and bomb load. Because such a wealth of detail is not apparent in a mere print, they use special spectacles to provide a 3-D image.

Before the advent of satellite observation, U-2 reconnaissance planes kept watch on military developments in the communist world. Aircraft reconnaissance continues today but the most daring flights came in the late 1950s and early 1960s when secret flights were made over the USSR to assess progress in missile development. The most famous incident was in May 1960 when a U-2 flown by Gary Powers fell to a Soviet ground-to-air missile on the approach to Sverdlovsk.

was perceived. Moreover, like any other bureaucracy, the PVO-Strany would naturally seek to augment its responsibilities to include space as well as air defense.

Soviet Space Troopers The Soviet Defense Minister Marshal Malinovskii said in February 1963 that the military had been given the task of 'combating an aggressor's modern means of nuclear attack and his attempt to reconnoitre our country from the air and from space'. A special ASAT section of the PVO-Strany was set up during 1964, designated the *Protivo Kosmicheskaya Oborona* (Against Cosmic Attack). The weapons of the PKO were described as 'special spaceships, satellite fighters and other flying apparatus armed with rockets and radio-electronic apparatus (. . .)'[9]

Despite such statements, the Soviets' actual ability to attack orbiting satellites before 1967 remains uncertain. In late 1960, they were believed to have the engines, boosters and electronics for an ASAT. Conversely, there were doubts that they had as yet put the elements together.[10] Others think that the USSR was able to intercept satellites by 1963.

In retrospect it appears that if the Soviets did have an ASAT during the 1960s, it was a direct-ascent system along the lines of the US Programs 505 and 437. The first likely candidate for such a Soviet system is the ABM-1 Galosh, a cone-shaped missile with three solid-fuel stages, about 65 ft (19.8 m) in length. Its warhead has a yield of 2 to 3 megatons and the Galosh's range is estimated to be at least 200 miles (322 km).

Exact details about the Galosh are difficult to determine since it is only displayed in a launch canister with only the four first-stage nozzles visible. The launch canister is perhaps lowered vertically into an underground silo from which the Galosh is fired, deploying its fins once clear.[11] If there was a Galosh ASAT, its probable site was the Saryshagan ABM range in Soviet Central Asia which has the necessary launchers, radars and trained personnel. Missiles and warheads could be maintained there on operationally ready status. Just as in the case of the US ASAT program employing the Nike-Zeus, fueling and electrical modifications would probably have to be made to adapt the Galosh for an ASAT function. So adapted, it ought to have performed at least as well as the Nike-Zeus ASAT and may have been operational by the mid-1960s.

The Soviets also had their own version of Program 437. During their 1962 test series, three nuclear weapons were exploded in space. These tests conducted over Soviet Central Asia[12,13] at an altitude of 250 miles (400 km), were similar to the US Starfish shot. The first was made on October 22, 1962. The Atomic Energy Commission estimated the yield as a few hundred kilotons. The second and third tests occurred on October 28 and November 1. Yields were described as 'intermediate' (ie, 20 kilotons to 1 megaton). One difference between the US Starfish shots and the USSR equivalents was that the radiation belts created by the Soviet bombs dispersed more rapidly. This was due to differences in the structure of the Earth's magnetic field between the equator and latitudes closer to the Pole.[14]

Cosmos 11 was launched by the Soviets two days before the first high-altitude test. Did it provide data about the effects of radiation on satellites, with a view to employing nuclear ASATs? The actual satellite-disabling applications of the Soviet high-altitude test system are hard to gauge. The effectiveness of any ASAT system depends on its ability to reach the altitude of the target satellite in the first place. If the 250-mile (400-km) altitude demonstrated in the nuclear tests was the best it could do, it had little or no advantage over a Galosh ASAT. Altitude capability depends of course on the launch vehicle. Just which vehicle the Soviets used is not known, but of the three candidates — the SS-3 Shyster, SS-4 Sandal and the SS-5 Skean — only the SS-5 has an altitude capability similar to the Thor.

It must be remembered that these are only possibilities: the Soviets did not necessarily follow the US pattern. Did the Soviets forego the more limited direct-ascent ASAT and concentrate all their efforts on an orbital interceptor? One thing is clear: unless the PKO was a paper organization — that is, a name and nothing more — the Soviets had either an equivalent of Program 505 and 437 or a pioneering ASAT system of some description.

SAINT and the Cosmos anti-satellites In the shape of Program 437 and its successors the US had developed a direct-ascent anti-satellite system. As we have seen this method (or profile) of satellite attack employs a ground-launched missile which directly intercepts the path of the target satellite. Its warhead explodes without delay, annihilating the target. The US adopted this profile (with modifications) in later ASAT research, but an entirely different approach — that of orbital rendezvous — was explored in the first place. Consequently, SAINT (Satellite Interceptor), the first US anti-satellite program and based on orbital rendezvous, anticipated the anti-satellite route chosen by the USSR.

The SAINT program of the late 1950s and early 1960s investigated the problems of two spacecraft rendezvousing in orbit. Although SAINT arose as a reply to the threat perceived from satellite weapons, its findings had applications in such diverse fields as space station assembly, re-supply and lunar landing. The Air Force had undertaken preliminary work on space rendezvous during the late 1950s, and the SAINT program was made public on September 24, 1960.[15] By the end of the year the Radio Corporation of America had won the development contract.

SAINT intended to use as much off-the-shelf hardware as possible to keep costs low. This included use of existing boosters and existing subsystems such as sensors.[16] Similarly, radar was a modified version of the Westinghouse system, designed for use in the Bomarc anti-aircraft missiles.[17]

The program had two phases, the first concerning the demonstration of rendezvous technology and the second satellite inspection. For the first phase, an Atlas D-Agena B comprised the launch vehicle with the SAINT making up the third stage. The vehicle itself was composed of a propulsion stage to conduct the orbital maneuvering and a payload of homing radar and television camera. Cone-shaped with the radar unit at the nose, the weight of the initial version was approximately 2400 lb (1089 kg).[18]

Launch was planned to take place from Cape Canaveral. Once the altitude of the target satellite — a radar reflector or large balloon satellite in a 400-nautical mile (741-km) high orbit — had been attained, the SAINT was to separate from the Agena B and begin final maneuvering. Initial tests with durations of between two and 48 hours, had relatively simple rendezvous profiles and cargos of sensor equipment to establish likely problem areas.[19,20]

While the initial research and development flights were made, work on advanced sensors including television and infrared cameras and nuclear radiation detectors would be underway.[21] It was crucial that the inspector system be able to determine with a high degree of confidence whether or not

SAINT – The Satellite Interceptor

Reconstruction of the satellite interceptor from a declassified source. The SAINT set out to inspect satellites by remote-control TV. It used complex orbital rendezvous, *jockeying alongside its target to obtain a close look. Like its cousin, the Discoverer 'spy in the sky', it was based on an Agena rocket stage.*

DISCOVERER

SAINT

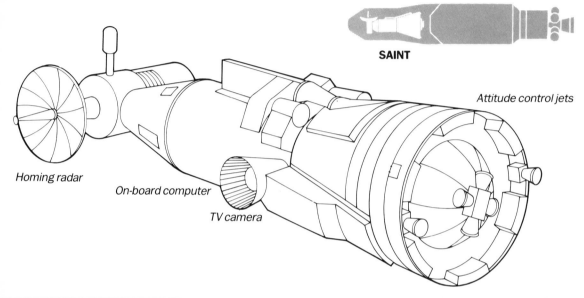

Attitude control jets

Homing radar

On-board computer

TV camera

The Agena B/SAINT assumes an elliptical orbit with its high point near the altitude of the target satellite's orbit. Then it ascends until its orbit resembles the target's. The SAINT then separates and begins to maneuver. When both satellites are in identical orbits SAINT inspects the target. The sequence is so complex that even a small error may jeopardize the whole mission.

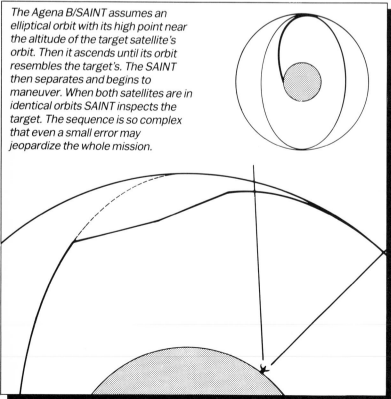

Above: *SAINT tried hard to use existing hardwear to keep costs low. It was accordingly modelled on the Agena Discoverer satellite shown here.*

the target satellite was carrying a nuclear weapon. After all, a positive report would trigger an international incident. This had to include the ability to discriminate between decoys and active satellites as well as between scientific and military payloads. It had to distinguish, for example, between a weather satellite and a reconnaissance satellite.[22] The emphasis was on inspection alone.

The second phase was scheduled for July 1965, with an Atlas-Centaur as the launch vehicle. Estimates suggested that the final version would weigh twice as much as the prototype. So-called 'dog leg' maneuvers were planned in which the SAINT would rendezvous with satellites on different orbital planes and 15 launches were envisioned. Total program cost was predicted as $1.2 billion.[23]

It was not to be, for SAINT faced problems in three areas. Off-the-shelf hardware in the first instance was proving inadequate to the task. Necessary modifications were so great that completely new development efforts were needed. Costs increased and the program fell behind schedule. Questions arose about reliability. Analysis by the Aerospace Corporation, the Air Force technical adviser and RCA indicated only a very small chance that a successful rendezvous would be accomplished during the four-flight demonstration series.

Reliability was also a problem because of the long and complex procedures which were necessary. First, the target satellite had to be launched successfully. Second, the SAINT launch had to take place within a very narrow time-span. Lastly, once in orbit, a complicated sequence of events had to be executed without error in order for a successful rendezvous to be accomplished.

Another problem was in the area of management. Observers of the program blamed the Air Force for not clearly defining its goals. A lack of firm management on the part of the Air Force Space System Division and RCA was suspected and sharp disagreements between the Air Force and the Aerospace Corporation over objectives and technical matters were rumoured.[24]

Doubts about the usefulness of the whole SAINT concept were the final blows. The sudden launching of a large number of satellites into low Earth-orbit would itself be a tip-off to an adversary. There may be no need, under these circumstances, for a detailed inspection of each satellite.

Another problem concerned decoys and disguises. If the US strategy was to launch a SAINT for every Soviet payload, then the Soviets could exhaust US defenses by simply putting up large numbers of decoys. The Soviets could also disguise their orbital nuclear weapons as scientific satellites. Moreover, simply looking at a satellite might not reveal its function. All that need be done to conceal a weapon was to encase it in a shroud, so that from outside it appeared as a large cylinder with the appropriate solar panels, horizon sensors and scientific instruments. If the satellite was believed to be hostile, why bother with the intermediate inspection step? Why not just destroy it directly? The most telling argument was that even if the SAINT could show that a satellite was armed, it could do nothing about this, since SAINT was itself unarmed.[25]

In December 1962, SAINT was cut back and eventually cancelled. Its brief, troubled life recalls Program 437. There was even opposition, on religious grounds, to the acronym which formed the program title. The inherent shortcomings of the project, coupled with those of its administration, were partly responsible. More important though was the obvious political requirement at this time for an operational ASAT to counter the orbital weapons threat perceived from the USSR. The direct-ascent profile employing the Nike-Zeus and Thor rockets provided this, rendering the developmental SAINT redundant.

The 'hot-metal kill'

The Soviet interceptor was in many respects markedly different from SAINT. Its fast flyby of a target was unlike SAINT's rendezvous and inspection. The crucial differences lay, however, in the fact that the Soviet model is meant to destroy a satellite. Practising what is known as the 'hot-metal kill', it is essentially a supergrenade which detonates in the vicinity of its target and produces a spherical cloud of shredded metal expanding evenly in all directions.

A high-explosive warhead circumvented problems with space legislation and the nuclear test ban treaty. It meant as well that the Soviet ASAT had a very narrow miss distance, having to pass within 3180 ft (1 km) of the target satellite for a successful 'kill'. This was because a conventional explosive warhead does not have the destructive effect (or 'kill radius') of a nuclear ASAT. As a result the Soviet ASAT had to meet taxing technical requirements, although its flight profile was not so complex as SAINT's.

When the US began to fly reconnaissance satellites in the early 1960s, there were worries that the USSR might attempt to destroy them. Accordingly the US kept a close watch on Soviet space activity, looking for any sign of a rendezvous capability.[26] The first hint was afforded by the Vostok 3 and 4 dual mission. On August 11, 1962 Vostok 3, with Andrian Nikolayev on board, was placed into a 146 by 113-mile (235 by 181-km) orbit, inclined 64.98°. The next day, Vostok 4 and its pilot Pavel Popovich went into a 147 by 112-mile (237 by 180-km) orbit.

Because of the similar flight paths and launch timing, the two spacecraft flew in close proximity as they circled the Earth. Their closest approach was 4 miles (6.5 km) shortly after Vostok 4 reached orbit, as against a planned separation of 3.1 miles (5 km). The two crewmen nonetheless could see each other's spacecraft. As the days passed, the slight differences in orbital characteristics of the two caused them to drift apart. The point is that the Vostok spacecraft lacked any rendezvous capability. A close approach depended on the precision of launch; once in orbit, they could do nothing to close the distance. This dual flight, and a similar one between Vostok 5 and 6, were apparently intended to impress the West with the capabilities of Soviet space technology.[27] If the Soviets could coordinate Vostok orbits in this way, they could do so with a target and interceptor.

The ASAT implications of the Vostok 3 and 4 missions were not overlooked. The British radio astronomer Sir Bernard Lovell pointed out that the Soviets had made a big step toward an ASAT with the Vostok 3 and 4 flights. A Soviet spokesman said that 'Soviet successes in space provide full assurance that in the distribution of the world's forces, there are all the means (necessary) for (. . .) liquidating any attempt at "war in space" or "through space".[28]

The 4-mile (6.5-km) separation between Vostok 3 and 4 was within the lethal radius of the MK 49 ASAT nuclear warhead and the Soviets had simultaneously demonstrated the ability to launch a rocket within the narrow time constraints necessary for a successful rendezvous.

The next goal was the vastly more difficult task of maneuvering in space. The first Soviet satellite capable of changing its orbit was the Polyot 1. After launch on November 1, 1963 it made extensive changes in both altitude and orbital inclination, going from an initial 368 by 210-mile (592 by 339-km) orbit to a final one of 893 by 215 miles (1437 by 343 km). Polyot 2, second and last in the series, was orbited on April 12, 1964, conducting orbital maneuvers like its predecessor. These flights appeared to be engineering test missions to aid the design of future Soviet spacecraft; most probably the Soyuz manned spacecraft.[29]

Seven years after Khrushchev had fulminated against reconnaissance satellites, the first flight test of Soviet orbital ASAT hardware was made on October 27, 1967. Over the course of the next five years, target and interceptor launchings of many kinds took place, flown under the Cosmos title. Analysis of these exercises points to a remarkable conclusion. The interceptor flights taken together demonstrated the USSR's ability to destroy photo-reconnaissance, ELINT, weather and Transit navigation satellites. The various altitudes and orbits of these satellites were approximated by the Cosmos targets and were all within range of the Cosmos interceptors as evidenced by the Soviet tests. One school of opinion holds that the early Soviet ASAT program was a countermeasure against any US attempt to orbit nuclear weapons. If so, the Soviet ASAT program arose in much the same fashion as that of the US; as a response to a threat.

F-1-m

SALT 1 and its after effects

After the 1971 Cosmos series was completed, Soviet ASAT operations came to a halt. Although the SALT 1 talks conducted at this time concerned the number of ICBMs, bombers and missile submarines permitted to each side and made no direct mention of ASATs, it may be that the SALT 1 treaty helped to curtail them. However, the Soviet interceptor tests left the US apprehensive, on the grounds that the reconnaissance satellites which would monitor compliance with the treaty could now be destroyed.

US negotiators specifically asked that a pledge of non-interference with reconnaissance satellites be included in the treaty as a result. The Soviet civilian negotiators, for their part, denied any knowledge of ASAT tests. When the US negotiators went on to provide information on each of the Soviet missions, the Soviet team fell silent. At least one of the US representatives concluded that the Soviets had not been lying but rather that they had been kept in the dark by the military.[30] The SALT 1 treaty signed in 1972 contained a clause prohibiting interference with 'national technical means of verification'; a euphemism for reconnaissance satellites. The ban on interference did not expressly prohibit the testing or possession of an ASAT.

At the same time, the Soviets were becoming aware of a space threat arguably more dangerous than orbital bombs or reconnaissance satellites. This menace emanated from civilian communication satellites. Since the Echo, Telstar, Relay, Syncom and Early Bird satellites of the early 1960s, still more impressive technical advances had been made and direct broadcast satellites were being designed in the early 1970s. These satellites were able to transmit directly to home TV sets and eliminated the need for large ground stations.

Soviet Foreign Minister Gromyko submitted a draft treaty on satellite broadcasting to the UN on August 8, 1972. It required that a nation should obtain permission before broadcasting directly to another country. Crucially, it went on specifically to ban the use of satellite broadcasting for interfering with intrastate conflicts, encroaching on human rights, purveying violence, horror, pornography or drug use, the subversion of local culture, tradition or language and misinforming the public — conditions which most TV programs would find it hard to meet. The treaty also contained an enforcement clause allowing the nation in question to use any means available to counter an unwanted satellite broadcast[31]. This direct broadcast clause of the satellite treaty would make the right to destroy satellites a part of international law. Soviet fears were premature, since to receive such broadcasts requires a dish antenna several feet in diameter and auxiliary electronic equipment; but the questions raised for the future as technology progresses are important ones.

Soviet anti-satellite tests (1967–72)

Cosmos 185
Oct 27, 1967 from Tyuratam by an F-1-m (the vehicle's first use).
Orbit profile: It assumed a low orbit of 339 × 230 miles (546 × 370 km), inclined at 64.1°. It then maneuvered into a 552 × 324 mile (888 × 522 km) orbit.
Purpose and comment: Perhaps this was an engineering test of the maneuvering system, or a target satellite which somehow failed before the interceptor was launched.

Cosmos 217
April 24, 1968 from Tyuratam by an F-1-m. (Subsequent target and interceptor launches followed this pattern.)
Orbit profile: The Soviets announced its orbit as 323 × 246 miles (520 × 396 km), inclined at 62.2°. Scattered debris was found by US tracking systems in a 163 × 71 mile (262 × 144 km) orbit.
Purpose and comment: The Soviets seem to have announced the intended final orbit. A malfunction apparently prevented the vehicle attaining this. Was this another target satellite which somehow failed before the interceptor was launched?

Cosmos 248
Oct 19, 1968.
Orbit profile: Assumed a 342 × 304 mile (551 × 490 km) orbit, inclined at 62.3°.
Purpose and comment: This was an unusually large target satellite, weighing c 10 000 lb (4536 kg). It presumably carried equipment to measure interceptor miss distance and sensors to monitor shrapnel pattern and explosion effects in space.

Cosmos 249
Oct 20, 1968 (a day after the target satellite).
Orbit profile: It assumed a 158 × 85 mile (254 × 136 km) orbit. After firing its on-board engine, it assumed an orbit of 1353 × 319 miles (2177 × 514 km).
Purpose and comment: After 2 orbits the interceptor or 'satellite killer' closed in on the target at a speed of c 1000 mph (1609 km/h). It swept past in a fast flyby, probably exercising the interceptor's on-board radar, and exploded. The official Soviet statement declared that 'the scientific research envisioned by the program has been fulfilled'. US military observers considered that, on the contrary, this ASAT mission was a failure. The interceptor's flyby and detonation had taken place outside the required 1-km 'kill radius'.

Cosmos 252
Nov 1, 1968.
Orbit profile: After a 155 × 87 mile (250 × 140 km) first orbit, it maneuvered into a 1350 × 334 mile (2172 × 538 km) attack orbit.
Purpose and comment: This 'satellite killer' apparently had the same objective as the previous one. Its fast flyby of the Cosmos 248 target which was still in orbit was this time inside the warhead's 'kill radius'. In a real interception, the target satellite would have been destroyed by the interceptor's conventional-explosive warhead.

Cosmos 291
Aug 6, 1969.
Orbit profile: Inclined 62.3° (a typical ASAT orbital characteristic).
Purpose and comment: An intriguing and inconclusive episode. After separating from its carrier rocket, the payload did nothing at all. It did not assume a circular orbit like the Cosmos 248 target of the year before, nor did it make any kind of maneuver. Was it yet another target for an interception which never took place?

Cosmos 373
Oct 20, 1970.
Orbit profile: US tracking stations detected Cosmos 373 in a 685 × 317 mile (1102 × 510 km) orbit, and watched it maneuver into a lower 343 × 305 mile (553 × 490 km) orbit.
Purpose and comment: A target satellite resembling the first target Cosmos 248 (launched on the latter's second anniversary).

Cosmos 374
Launched three days after the Cosmos 373 target.
Orbit profile: It first assumed a 654 × 329 mile (1052 × 529 km) orbit, then fired its on-board engine to enter a 1338 × 333 mile (2153 × 536 km) orbit.
Purpose and comment: The interceptor detached itself from the carrier rocket once final orbit was attained. The interceptor module then hurtled past the target outside the necessary 'kill radius' and exploded. An apparent failure.

Cosmos 375
Oct 30, 1970.
Orbit profile: First entering an eccentric orbit of 621 × 311 miles (1000 × 500 km), it shortly afterwards assumed a 1345 × 333 mile (2164 × 538 km) orbit.
Purpose and comment: Like its forerunner Cosmos 374, the interceptor module again detached itself from the carrier rocket and closed in. Unlike the former, it passed within 1 km of the target before detonating.

> **ASAT flight tests in 1968 and 1970: the differences**
> It seems that these two early test series were designed to establish a basic anti-satellite system. Soviet efforts most likely centered on overcoming rendezvous problems and monitoring high-explosive detonations in space so as to refine the system further. Both the 1968 and the 1970 series showed much the same profile; each time making 2 circuits of the Earth before attacking as the interceptors neared the low points of their orbits. Orbital behaviour also differed. The 1968 ASAT interceptors went into low elliptical orbits before maneuvering into higher and even more elliptical ones. In contrast, the 1970 ASAT interceptors experimented with a new attack profile: assuming first a high elliptical orbit, then doubling the apogee but leaving the perigee nearly unchanged.

Cosmos 394
Feb 9, 1971 — evidently launched by a C-1 booster from Plesetsk (an entirely new departure).
Orbit profile: Assumed a 385 × 357 mile (619 × 574 km) orbit, inclined 65.9°.
Purpose and comment: A target satellite, lighter than previous ones at 1500 lb (680 kg), with an orbit higher than earlier targets and a new inclination. This target also broke the previous pattern in that it was used for only one interception.

Cosmos 397
Feb 25, 1971 by an F-1-m.
Orbit profile: After entering an elliptical 381 × 89 mile (613 × 144 km) orbit, the interceptor module with warhead separated from the carrier rocket and maneuvered into a 1440 × 368 mile (2317 × 593 km) final orbit.
Purpose and comment: A 'satellite killer' which successfully hurtled past its target within 'kill' range before destroying itself.

Cosmos 400
Mar 19, 1971.
Orbit profile: It entered a 631 × 618 mile (1016 × 995 km) orbit; the highest yet for a target satellite.

Cosmos 404
Apr 4, 1971.
Orbit profile: Its initial orbit ranged from 393 to 92 miles (632 to 148 km). After separating from the carrier rocket, the interceptor module maneuvered into a 627 × 504 mile (1009 × 811 km) orbit.
Purpose and comment: This flight was innovatory in that the flight path of the predator practically matched the victim's (ie, Cosmos 400). The flyby was leisurely and their relative velocities quite low; an interesting departure from the original pattern. But speculation that this ASAT had an 'inspector' role is quashed by the fact that despite its relatively low speed it spent only a few seconds within 5 miles (8 km) of the target. This is probably too little time for an inspection. After the flyby, the interceptor retrofired and adopted a lower orbit: 496 × 105 miles (799 × 169 km), before re-entering and burning up over the Pacific.

Cosmos 459
Nov 29, 1971.
Orbit profile: It assumed a very low orbit, 172 × 140 miles (277 × 226 km).
Purpose and comment: The orbit in this case — the lowest orbit of any target so far — suggests that it was intended to simulate a reconnaissance satellite.

Cosmos 462
Dec 3, 1971.
Orbit profile: The initial orbit was 970 × 89 miles (1561 × 143 km); a more elliptical orbit than any previous vehicle in this class. It then made minor corrections, adopting a 1143 × 147 mile (1840 × 237 km) orbit.
Purpose and comment: An interceptor which made a successful fast flyby before detonating. Like the other flights made in 1971, this interception was an ambitious one. The satellite 'kills' of this year demonstrated the USSR's ability to eliminate photo-reconnaissance, ELINT, weather and Transit navigation satellites.

Project SPIKE

The years passed and no new interceptor tests were made. It seemed that the Soviets had abandoned ASATs. The US Program 437 was phased down and interest in new ASAT systems was sporadic. None were developed. One proposal made during this period, however, had great significance. Project SPIKE began with a suggestion by Colonel Hugh D Dow of ADC Headquarters that an air-launched ASAT could be built using aircraft and off-the-shelf rockets.[32] SPIKE employed an F-106 carrying a Standard anti-radar missile with a small second stage and terminal homing vehicle. Satellite tracking data was relayed via existing ground-to-air transmitter sites to the F-106's on-board computer. This information would be used to position the aircraft for launch. The launch altitude would be about 35 000 ft (10 668 m).

SPIKE aimed to destroy satellites in low-altitude orbits. Using existing F-106 bases, satellites could be intercepted before their first pass over the continental United States. Only minor modifications need be made to the aircraft and they would not interfere with the F-106's regular missions.[33] The emphasis was on acquiring a low-cost, fast-reaction capability to destroy satellites; SPIKE was not an experiment or a retaliatory show of force like Programs 505 and 437.

The most technically demanding part was the terminal homing vehicle. This unit would include the primary seeker, horizon sensors and a small on-board computer. Wrapped around it were multiple rings of small solid-fuel rockets. The idea was that after separation the terminal-homing vehicle would begin to rotate, and the data from the horizon-roll sensors and primary seeker meanwhile fed to the on-board computer. It was then up to the computer to give steering commands and fire specific rockets to maneuver the vehicle.[34] The payload might be a small nuclear or non-nuclear warhead, or a photographic package. When the initial studies were completed, a round of high-level briefings began. The ADC and Systems Command were ordered jointly to determine Project SPIKE's priority and to include it in pending ASAT systems analyses. General Seth J McKee, Commander-in-Chief of NORAD was briefed on SPIKE. He was 'impressed by the potential utility and flexibility of such a system'.[35] Technical studies continued and a decision on SPIKE was postponed. The results when they came showed that the project was feasible, with modifications.

A development program was mapped out. It included a six-shot interception program at Eglin-Tyndall AFB, Florida. The target satellites were intended to realistically simulate the characteristics of specific Soviet and Chinese satellites. At the same time, doubts about the project were beginning to surface. Joe C Jones, Deputy Assistant Secretary of the Air Force for Research and Development, framed some pertinent questions. Why did the ADC want to do it? Could a modified Standard missile really reach useful altitudes? Was the concept too sophisticated to realize without major technical advances? Doubts lingered.[36]

The final question was the most important. The briefing team stressed the use of available hardware and technology, implying that development was reasonably simple. This was SPIKE's primary advantage. Assistant Secretary Jones, on the other hand, felt that the terminal homing vehicle would require a major development program, and so SPIKE was not given the go-ahead. However, although they did not amount to a full-scale development effort, studies were made in the wake of Project SPIKE which provided the foundation for a future US plane-launched ASAT.

This period also saw disquieting developments in the USSR. Cosmos 521 was launched by a C-1 booster from Plesetsk on September 29, 1972 to enter a 640 by 605-mile (1030 by 973-km) orbit, inclined at 65.8°, resembling the Cosmos 400 target of the previous year. Was Cosmos 521 the target for an interception later abandoned? Three years afterwards, a C-1 booster from Plesetsk on July 24, 1975 put Cosmos 752 into an orbit with an inclination of 65.9°. It assumed a 327 by 298-mile (526 by 480-km) orbit — unlike that of previous Soviet targets. Apparently the first flight of a new species of ASAT-related satellite, Cosmos 752 presumably carried special on-board monitors and exercised ground radar (such cargos are known in technical circles as diagnostic and calibration payloads).

The Chinese connection Seven months later the Soviets resumed ASAT testing with the launching of a Cosmos 803 target by a C-1 from Plesetsk on February 12, 1976. It was the start of another test series which, in the next flight, saw an entirely new profile — with alarming consequences for US military observers. The interceptor Cosmos 814 ushered in a new era of anti-satellite warfare when it lifted off on April 13, 1976. As the first appearance of a disturbing new launch profile (which later came to be called 'Pop Up') it deserves consideration in detail. Cosmos 803, still in orbit from the previous test, formed the target once again and the interceptor was launched some four minutes after the target vehicle had passed over the launch site. NORAD tracking data showed the Cosmos 814 interceptor going into an initial 297 by 72-mile (479 by 116-km) orbit, which was surprising since this was considerably lower than the target's 385 by 340-mile (620 by 548-km) orbit. This lower orbit meant that Cosmos 814 soon gained on its target.[37] Once it had caught up in this way, Cosmos 814 fired its on-board engine, assumed an elliptical orbit and made a fast flyby.

C-1 COSMOS

The Soviet Cosmos ASATs explored avenues other than the direct-ascent method used by the US. Concentrating on orbital rendezvous, they employed a 'parking orbit' in which the interceptor is put into the orbit plane of the target irrespective of the target's position. The interceptor quickly gains on the target (as it has a shorter orbital period) and maneuvers to attack. Three variations were played on this theme.

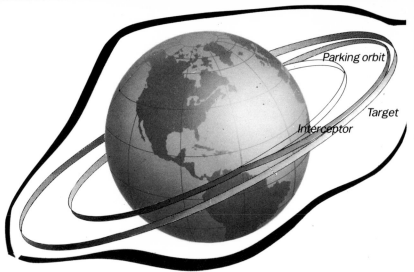

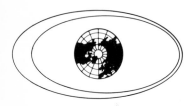

1 In the co-orbit technique, the interceptor keeps pace with the target, finally put into coincidence with it by thrust impulses.

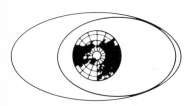

2 In the fast flyby, the interceptor gains more quickly by its chasing technique.

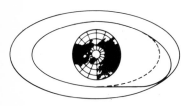

3 In the very flexible 'Pop Up' maneuver, the interceptor 'pops up' into the target's orbit to make its 'kill'.

The whole affair, from launch to interception, had taken approximately 42 minutes. Cosmos 814 went on to re-enter and burn up, but its implications were etched indelibly on the minds of Western military observers.

This appearance of the 'Pop Up' profile introduced an anti-satellite system which required less than one orbit from launch to interception. It abandoned the laborious maneuvering of earlier attack profiles and provided an authentic fast-reaction capability. If everything was ready at launch, the Soviets could destroy US satellites while they were out of US tracking range. Hitherto, the maneuvering of an ASAT interceptor served as warning that an attack was underway. With the 'Pop Up' profile, a US satellite could go beyond tracking range and be scattered debris when it returned; the victim of a convenient 'accident'.[38] Nor was this all. In the event of a full-scale conflict, Big Bird and KH-11 reconnaissance satellites (which use an on-board engine to prevent premature orbital decay) could have taken evasive action. Against a fast-reaction system, any action would be ineffective.

Washington officials were obviously becoming concerned about the renewed Soviet ASAT activity. US military satellites were considered to be endangered and a US counterforce deemed necessary. Consequently, the outgoing Ford Administration approved a new ASAT development effort. Two days later, Jimmy Carter was inaugurated as President, and confirmed the decision, albeit that the program was to be small-scale.[39] Simultaneously, an agreement with the Soviets banning space weapons was to be actively sought.

Nine months later on the twentieth anniversary of Sputnik 1, Defense Secretary Harold Brown briefed the press. He announced that the Soviet ASAT was operational, that it could destroy certain US satellites, and described it as 'somewhat troubling'. In marked contrast to the operational Soviet system, the new US ASAT was still in the teething stage, the Vought Corporation having only recently been awarded the design contract. Despite the existence of the Soviet ASAT, Brown expressed his hope to 'keep space from becoming an arena of active conflict'.[40]

During the 1976-1977 test series, the new 'Pop Up' technique was tested in a variety of circumstances. First, against a target in a medium-altitude circular orbit; then in a high elliptical orbit. Next a target was placed in a low elliptical orbit and finally into a high circular one. Each situation put different demands on the interceptor. However, the series did not demonstrate an improved altitude capability. The Navstar navigation satellites (in their 12 500-mile |20 116-km| high orbit), geosynchronous communications and early warning satellites were still out of reach.[41]

Western analysts came to the conclusion in the late 1970s that concern over the Chinese space program was playing a large part in shaping the new Soviet ASAT effort. Several items of evidence supported this belief. In the first place, to paralyze completely US space operations the Soviets would have to employ a large number of ASATs. On-orbit spares were available to replace disabled US communications satellites and U2 and SR-71 aircraft could take the place of reconnaissance satellites.

The Chinese space program, on the other hand, involved only a very limited number of satellites confined to low Earth-orbit. These could be easily dealt with by Soviet interceptors. The actual timing of the Soviet tests also tended to support the China theory. Soviet ASAT flights were made soon after the first two Chinese satellites were orbited in April 1970 and March 1971. No further Chinese launches or Soviet ASAT tests were made for the next few years. Then in 1975 the Chinese made three launches — in July, November and December — which included the first successful recovery of a re-entry capsule. This is an essential prerequisite for flying a film-returning reconnaissance satellite. Two months later, the Soviets resumed ASAT testing. If this body of evidence were not enough, there was the matter of similarities between the orbits of the Soviet targets and those of Chinese satellites.[42] Finally, reports in 1981 that a ground monitoring station had been built in China also seem to bear the West's theory out.

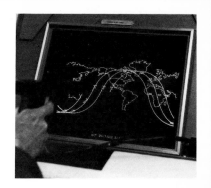

Within the Cheyenne Mountain complex officers of NORAD keep watch over the world's ever-growing population of artificial satellites. A computerized catalogue is kept of every orbiting object, working or defunct, down to small metal scraps. In the data banks at the time of writing are more than 13 000 items of space debris dating from October 1957. Many of these have long since disappeared from orbit as contact with the thin upper air causes them to spiral down to destruction, but some 6000 man-made objects remain. They range from small metal scraps the size of a saucer to satellites weighing several tons. At any time the ground track of a satellite can be reproduced on a video screen. This one shows the path of China's second satellite.

Carter's bid for peace

Defense Secretary Brown's hope that space would remain free of hostilities reflected a major goal of the Carter Administration. President Jimmy Carter had been elected on a platform of arms control and reduced military spending. One of Carter's aides had been quoted as saying 'He doesn't even want a peashooter out there'.[43] Although President Carter reaffirmed the Ford Administration's decision to begin US

ASAT development, it was with reduced emphasis. The program was to be both a carrot and a stick to entice the Soviets to begin ASAT talks.

The Carter Administration began the groundwork soon after Inauguration. At a press conference on March 9, 1977 Carter stated:

'I have proposed both directly and indirectly, to the Soviet Union, publicly and privately that we try to identify those items on which there is relative agreement (. . .) I have, for instance, suggested that we forego the opportunity to arm satellite bodies and also forego the opportunity to destroy observational satellites.'[44,45]

The Soviets privately expressed a willingness to discuss such a ban. Although the primary emphasis was on the SALT II treaty, work continued through the summer on the ASAT talks. By the autumn of 1977, various draft proposals were being formulated for subsequent submission to the Soviets, who continued to express interest in the talks.[46]

Within the US Government, however, there was considerable disagreement. The Defense Department wanted to delay the start of the talks until it had completed a space policy study for the National Security Council. There was also concern about the impact of the various negotiating options on the US ASAT development effort and the problem of verifying such a treaty once in force — an area that received intensive study.[47] The State Department and the Arms Control and Disarmament Agency wanted the talks to begin without delay. In the meantime, the State Department agreed that US ASAT development could continue pending an agreement.

Another area of contention was exactly what the treaty would cover: a complete ban on ASAT deployment; on development; or simply a halt to testing such systems?[48] An inner agency working group was organized to determine policy and settle the differences. A similar procedure had been used for the SALT treaty.[49] During the early weeks of 1978, this Cabinet-level group, which included Defense Secretary Brown and Secretary of State Cyrus Vance, attempted to tackle the problems but by mid-March differences still remained.

Despite this, President Carter decided to begin the formal negotiations for an ASAT treaty. Secretary Vance notified Soviet Ambassador Anatoli F Dobrynin that the US wished to begin the talks in April — probably in Geneva, Switzerland. It was hoped that the internal differences could be settled by the time the talks had reached an important stage.

The Soviets publicly indicated their willingness to start on March 31, 1978.[50] As the preliminary work was underway, Cosmos 1006 was launched into a circular 253 by 237-mile (407 by 382-km) orbit: it appears to have been a diagnostic and calibration payload. One week later (on May

19, 1978), they provided an unambiguous reminder of their lead in ASATs by putting the Cosmos 1009 interceptor into a 590 by 90-mile (950 by 145-km) orbit. It was then maneuvered into first an 860 by 600-mile (1384 by 965-km) orbit, and finally into the 625 by 334-mile (1006 by 537-km) attack orbit, passing the Cosmos 967 target at the start of the interceptor's third orbit. The attack was successful. Cosmos 1009 then made a de-orbit burn and re-entered over the Western Pacific.[51]

The Soviet and US groups set out their basic positions and understandings in preparation for the talks. At an early stage both agreed that outside issues, such as strategic weapons systems or the neutron warhead, would not enter the ASAT discussions. The Soviets then took a position that the US had feared, making it plain that they considered the Space Shuttle a potential ASAT. Their statements were phrased in loose terms. The Shuttle was referred to as a 'manned system of dual use' and the US was not sure if the Soviets were objecting to certain Shuttle capabilities or the entire program. This naturally made a response difficult.[52]

The talks began on June 8, 1978 in Helsinki, Finland, the US delegation being headed initially by Paul C Warnke, Director of the Arms Control and Disarmament Agency. He

The Space Shuttle Challenger. Soviet objections to the Shuttle were an obstacle to President Carter's initiatives in 1977–78. They feared that it would be able to pluck their satellites from space with impunity and consequently continued their ASAT tests. This fear was given more substance when Challenger actually proved its capability to launch and retrieve satellites on June 22, 1983.

Soviet anti-satellite tests (1976–77)

Cosmos 803
Feb 12, 1976 from Plesetsk by a C-1.
Orbit profile It assumed an orbit of 388 × 344 miles (624 × 554 km), inclined 66°.
Purpose and comment A target satellite for the next launch.

Cosmos 804
Feb 16, 1976 from Tyuratam by an F-1-m booster.
Orbit profile The initial orbit was 436 × 92 miles (701 × 148 km). During its first revolution the interceptor maneuvered into a 383 × 548 mile (617 × 560 km) orbit. In the next 3 hours it changed again into a 385 × 328 mile (628 × 528 km) orbit, close to that of the Cosmos 803 target. During its fifth orbit, the Cosmos 804 interceptor made a slow flyby of the target, with the point of closest approach occurring about 8 hours after launch, just south of Havana, Cuba. After this interception, Cosmos 804 maneuvered again into a 557 × 142 mile (928 × 228 km) orbit before re-entering over the Pacific.
Purpose and comment Cosmos 804 behaved somewhat like the Cosmos 404 interceptor in 1971. Was this flight testing an inspection system?

Cosmos 814
April 13, 1976 from Tyuratam by an F-1-m, lift off taking place 4 min after the Cosmos 803 target (in orbit from the previous test) had passed over the launch site.
Orbit profile NORAD tracking data showed the Cosmos 814 interceptor taking an initial orbit of 297 × 72 miles (479 × 116 km), considerably lower than the target's. This meant that the interceptor gained swiftly on its quarry.
Purpose and comment Firing its on-board engine, Cosmos 814 went into an elliptical orbit. About 42 min after launch made a fast flyby of the Cosmos 803 target, within the 1-km limit of a successful interception. Cosmos 814 went on to re-enter and burn up. This new profile — later christened 'Pop Up' — required less than one orbit from launch to interception. It did away with the extensive maneuvering of earlier interceptors, and was an authentic quick reaction anti-satellite.

Cosmos 816
April 28, 1976 from Plesetsk by a C-1 (setting the pattern for subsequent launches).
Orbit profile It assumed an orbit of 326 × 299 miles (525 × 482 km).
Purpose and comment A diagnostic and calibration satellite, which ultimately blew itself up. (This is standard Soviet procedure for disposing of sensitive equipment.)

Cosmos 839
July 9, 1976.
Orbit profile It assumed an orbit of 1305 × 611 miles (2102 × 984 km).
Purpose and comment Cosmos 839 was put into the highest orbit of any target thus far.

Cosmos 843
July 21, 1976
Orbit profile It had an initial orbit of 223 × 93 miles (360 × 149 km), and re-entered soon after launch.
Purpose and comment A very unusual and thought-provoking flight profile. Did the planned Pop Up maneuver of this interceptor misfire? Or did the Pop Up occur over a remote area of the Soviet Union? Or was it even a successful interception with such a brief Pop Up that it escaped the notice of Western tracking stations?

Cosmos 880
Dec 9, 1976.
Orbit profile It entered a circular orbit of 384 × 348 miles (618 × 560 km), with an inclination of 66°.
Purpose and comment The orbit of this target satellite closely resembled that of Cosmos 803.

Cosmos 885
Dec 17, 1976.
Orbit profile It went into standard 319 × 292 mile (513 × 470 km) orbit.
Purpose and comment Snortly after going into orbit, Cosmos 885 exploded. It was not, however, the expected interceptor but rather a diagnostic and calibration payload.

Cosmos 886
Dec 27, 1976.
Orbit profile It entered an orbit of 355 × 82 miles (571 × 132 km) and then maneuvered into a 796 × 341 mile (1281 × 549 km) orbit. About 100 min after launch, it made a fast flyby of the Cosmos 880 target. Within 3 hours, it maneuvered into a 1427 by 378 mile (2297 × 608 km) orbit and then blew itself into fragments.
Purpose and comment After extensive maneuvering, Cosmos 886 was the only one of this series of interceptors to explode.

Cosmos 891
Feb 2, 1977.
Orbit profile It was placed in a 323 × 289 mile (518 × 466 km) orbit.
Purpose and comment The first anti-satellite related launch of the new year, Cosmos 891 was a diagnostic and calibration payload.

Cosmos 909
May 19, 1977.
Orbit profile It went into a 1306 × 618 mile (2102 × 996 km) orbit.
Orbit profile A target for the Cosmos 910 interceptor launched 4 days later.

Cosmos 910
May 23, 1977 from Tyuratam by an F-1-m.
Orbit profile It was sent into a 314 × 93 mile (506 × 149 km) orbit and re-entered before completing a full orbit.
Purpose and comment The total lifetime of this interceptor was only 70 min. It is reported that only one US radar, the Shemya Island tracking station in the Aleutians, registered Cosmos 910. This interception resembled the Cosmos 839–843 flights of the previous year. The targets had nearly identical orbits, and both flights seem to have been failures.

Cosmos 918
June 17, 1977.
Orbit profile First detected in a 122 × 77 mile (197 × 124 km) orbit. An on-board engine then sent it into a highly elliptical orbit.
Purpose and comment Piecing together data from the Shemya radar, it appears that the Cosmos 918 interceptor climbed toward the Cosmos 909 target in its 1307 × 618 mile (2103 × 994 km) orbit, and made a successful flyby. It went on to hit the Earth's atmosphere and burned up before completing a full orbit.

Cosmos 933
July 22, 1977.
Orbit profile An uncommonly low orbit for such a flight at 260 × 239 miles (418 × 385 km).
Purpose and comment Another diagnostic and calibration satellite, its exact purpose uncertain.

Cosmos 959
Oct 21 by a C-1 from Plesetsk.
Orbit profile It assumed a low elliptical orbit of 509 × 98 miles (820 × 158 km).
Purpose and comment A target.

Cosmos 961
Oct 28, 1977 by an F-1-m from Tyuratam.
Orbit profile Its initial orbit was 187 × 78 miles (302 × 125 km). It then maneuvered into an orbit of 882 × 167 miles (1421 × 269 km) – the attack orbit.
Purpose and comment Cosmos 961 employed the Pop Up profile and passed by the Cosmos 959 target within the required 1-km 'kill radius'. It then re-entered and burned up over the Pacific, giving rise to numerous UFO reports in Japanese newspapers.

Cosmos 967
Dec 13, 1977.
Orbit profile It was launched into a high circular orbit of 615 × 606 miles (991 × 997 km).
Purpose and comment A target.

Cosmos 970
Dec 21, 1977.
Orbit profile During its first revolution, this vehicle went from a 534 × 90 mile (854 × 144 km) orbit to an attack trajectory of 712 × 588 miles (1139 × 940 km).
Purpose and comment It flew past the Cosmos 967 target, outside the 1-km 'kill radius' and detonated its high explosive warhead. US radars detected about 30 fragments.

was replaced later in the week by Robert W Buchheim, the ACDA Deputy Associate Director for nuclear weapons and advanced technology. Representing the Soviets was Oleg Khlestov, head of the Treaty and Legal Affairs Division of the Soviet Foreign Ministry.

Initial discussions were meant to lay the foundation for formal negotiations later. Warnke, after a day of sparring, had an after-hours talk with Khlestov in which he warned his opposite number of the consequences if agreement was not reached. The US would be forced to build an ASAT superior to the Soviets'. He further warned that an incident in space might rapidly escalate. The only way to fend this off was to reach a basic agreement that ASATs were undesirable, then to formalize it in a treaty banning their development and testing. Khlestov responded by again condemning the Shuttle.[53]

On a more positive note, in the talks themselves the Soviets asked numerous questions about the US position, without however directly admitting that they had an operational ASAT. The talks concluded on June 16. Despite the continued objections to the Shuttle, the US negotiators felt that progress had been made.[54] A joint statement issued the next day declared that each side had developed a better understanding of the other's views, and talks were to be resumed later since both sides needed time to prepare more formal presentations.[55]

The second session of the ASAT talks was held at Bern, Switzerland between January 23 and February 16, 1979. Further progress was made during the preliminary sessions and White House officials believed that the next round might finalize the ASAT treaty — perhaps in time for a Carter-Brezhnev summit.[56,57] A greater degree of Soviet interest was one cause for optimism.

The third series of talks, which opened on April 23, 1979 in Vienna, was expected to last four or five weeks.[58] The US submitted a three-part proposal, calling for an end to ASAT tests, the dismantling of the Soviet's ASAT system and mutual verification of the other's good intentions.

By the end of May, the Soviets had disappointed hopes for an early agreement by demanding that a proposed moratorium on ASAT testing include the Space Shuttle. The phrasing of the treaty as proposed by the Soviets ruled out any spacecraft with a basic maneuvering capability which might conceivably be used against other satellites. The US flatly refused to consider either a ban or a delay where the Shuttle was concerned. The US point of view insisted that ASATs alone should be banned; this the Soviets resisted.

There were sharp differences as well over the specific language of the pledge not to 'damage, destroy or displace' satellites, the Soviets suggesting that only satellites belonging to the US and USSR be so protected. The US wanted the

The Northern Cosmodrome. A NASA Landsat Earth resources satellite looks down upon one of the USSR's most secret places – the cosmodrome near Plesetsk, south of Archangel. Here the Soviets launch Cosmos satellites into high inclination orbits, including targets for 'hunter killer' satellites launched from Baikonur. There are also launch facilities for medium-range and intercontinental ballistic missiles.

President Carter and Leonid Brezhnev at the conclusion of SALT 2 negotiations in Vienna 1978. Although the treaty was never ratified, the superpowers have broadly kept to its guidelines in missile deployments. Treaties governing military activity in outer space have been encouraged by the USSR. The Soviet-sponsored treaty banning nuclear weapon tests in the atmosphere, in outer space and under water, came into force in 1963. The Treaty on Principles Governing the Activities of States in the Exploration and Use of Outer Space, including the Moon and other Celestial Bodies (signed in 1967) provided for partial demilitarization of outer space and banned the use of the Moon and other celestial bodies for military purposes. In December 1981, at the instigation of the USSR, the UN General Assembly called for negotiations to draft an international treaty banning the deployment of all types of weapons in outer space. The Soviets have tabled the draft for such a treaty.

agreement to cover *all* satellites, including those launched by NATO and private organizations. The Soviets also wanted to retain the right to attack a satellite if it committed a 'hostile act', a condition which would effectively make the ASAT treaty meaningless. Pressed to define 'hostile act', the Soviets gave as an example a direct broadcast satellite.

Enter the F-15 anti-satellite

In the view of the Western delegates, the Soviets were deeply concerned about the Chinese space program. They did not wish to surrender an ASAT capability that might make a critical difference in any future Sino-Soviet war. The Chinese factor was a very difficult obstacle for an ASAT treaty since it was not under the control of either side. As the summit between Carter and Brezhnev approached, ASAT talks ground to a halt. Both sides waited for additional instructions from their governments which never came. Yet despite the failure of the talks, neither party wanted to be the first to go home.[59,60] By the end of June, the talks were finished — the plan was to resume the next round in the autumn.

This fourth round was in fact never held. On November 4, 1979 a mob of Iranians climbed over the wall of the US embassy in Teheran and took the diplomats hostage. A month later, the Soviet Army invaded Afghanistan. President Carter withdrew the SALT II treaty and broke off virtually all discussions with the Soviets. Detente was dead.

Despite the optimistic statements of government spokesmen, the US was in a difficult position with the ASAT talks. Essentially, the US had only two bargaining chips: the F-15 ASAT and the Shuttle. It was further hoped that the USSR would not want to engage in an ASAT race with the 'technically superior' US. An agreement banning ASATs would forestall this. The Soviets apparently did not see things in the same way. From their point of view, the F-15-launched ASAT was a paper weapon, one that the Carter Administration was not serious about.[61]

Throughout the ASAT talks, the Soviets had refrained from making orbital interceptions. At the same time, their ASAT program was not completely inactive. They continued to launch diagnostic and calibration payloads — Cosmos 1075 on February 8, 1979, Cosmos 1146 on December 5, 1979 and Cosmos 1169 on March 27, 1980.

Even after several years of such flights, their exact role in the Soviet ASAT program is unclear. They were launched seemingly at random; months before the interception tests, between the launchings of the target and interceptor and shortly after a test was over. If the diagnostic and calibration satellites were an integral part of the interceptor tests, a pattern ought to emerge. The lack of any such pattern, along with their regular two flights per year schedule and their stereotypical orbit, indicates that they were a separate part of the program.

When the ASAT talks collapsed, the design of the F-15-launched interceptor had been finalized. Its highly sophisticated 'satellite-killing' cargo was to be the Vought Miniature Vehicle (MV), referred to as Project 1005 (a play on the Roman numerals). As its name implies, it is a small object, 12 in by 13 in (4.7 by 5.1 cm). It employs an infrared sensor which looks out through 8 small telescopes to provide target information. This guarantees accurate data and prevents the Miniature Vehicle from 'attacking' stars. It spins at 20 revolutions per second, which not only keeps it stable but assists the 8 telescopes in acquiring and locking onto the target. Once the target is acquired, the Miniature Vehicle separates from the second stage.

The outer shell of the Miniature Vehicle is composed of 56 small cylinders of solid rocket-propellent, the nozzles of which point out to the side. When fired, under control of the guidance system, they move the vehicle bodily to keep it on a collision course. The rockets are fast-burning so as not to upset the spin stabilization. The guidance task of firing the correct rocket at the proper time is a major one requiring extremely sophisticated electronics, and timing is of the essence because of the vehicle's fast spin rate.

After the Miniature Vehicle's course has been corrected, a counterfiring stops the lateral drift. To achieve this accuracy, use is made of a laser-gyro which acts as a clock enabling the on-board computer to determine which rocket to fire. The computer also keeps track of which rockets have fired — they are single-shot only — and allows the Miniature Vehicle to rotate past the spent rockets. Additional rockets are used to prevent the Miniature Vehicle developing 'wobble' due to the firings.

The Miniature Vehicle destroys by direct collision with the target at 45 000 ft per second (13 716 mps). The energy of such an impact is akin to hitting a satellite with a shell from a battleship's main gun. Ironically, a direct collision at such high closing velocity is simpler than fuzing and exploding a warhead. The guidance system has no moving parts.[62,63]

This F-15-launched anti-satellite missile is a two-stage solid-fuel rocket 17.75 ft (5.4 m) long and weighing 2632 lb (1194 kg). The first stage 17.6 in (6.9 cm) in diameter is based on the Boeing Short Range Attack Missile (SRAM). At the base are two small fixed fins and three large movable fins which control the vehicle during atmospheric ascent. The second stage has an Altair III rocket motor of the kind used as the fourth stage on the Scout launch vehicle. It is specially strengthened for its anti-satellite role and is fitted with small hydrazine thrusters for attitude control. The second stage is 19.76 in (7.8 cm) in diameter. At its forward end is the Miniature Vehicle, with its spin table and subsystems (such as the inertial reference unit, computer and cryogenic tank for cooling the infrared sensor). An inertial guidance unit provides control during powered flight until a specific point in space is reached. At this point, the Miniature Vehicle begins to search for its target. After second-stage burnout, it spins up and the target satellite is acquired.

The F-15 launch aircraft itself requires certain modifications. An electronic package replaces the 20 mm ammunition container. There are wiring changes and a special centerline pylon which includes a microprocessor, a communications line between the missile and aircraft, a back-up battery, electrical connections and a gas generator ejection system.[64] The pilot's launch duties are minimal as he receives steering commands via the cockpit head up display.

Advantages of the F-15 ASAT

For most attack profiles, the ASAT will be launched while the F-15 is in subsonic, straight and level flight. For satellites in higher orbits, a supersonic climb is used. This adds speed to the ASAT and avoids the need for a sharp pull up which might overstress the missile. The launch is automatic with a 10 to 15-second window.[65] Two squadrons of ASAT F-15s will be established: one based at Langley AFB, Virginia and the other at McChord AFB in Washington. These locations were chosen because of the orbital inclinations of Soviet satellites, F-15 basing plans and the requirement that booster debris fall into the ocean.

The F-15 ASAT weapon is part of a closely coordinated system involving not only the aircraft and missiles but the ground-based attack and command structure as well. This is provided by the Space Defense Operations Center, a separate control room within NORAD's underground Cheyenne

Mountain headquarters. ASAT attacks are commanded by the Senior Weapons Officer who is in communication with the ASAT Mission Operations Center (also in Cheyenne Mountain), providing timing and attack information. In addition to coordinating ASAT attacks, the Space Defense Operations Center monitors the status of US satellites and ground stations. It is vigilant for any activity that would indicate an attack — a Soviet ASAT launch for instance, or radio interference. Warnings can be relayed immediately to spacecraft operators and national authorities.[66]

The F-15 ASAT has a number of advantages over a more conventional system. An F-15 can be flown to wherever necessary to accomplish an interception. A fixed-based ASAT, dependent on a large rocket, lacks such flexibility. As long as there are F-15s, ASAT missiles, supplies and means of providing target data, this highly mobile ASAT can continue to function whereas a fixed-base ASAT (such as that on Johnston Island) would be a candidate for attack during the early stages of an escalating war. It is economically feasible to build enough of the weapons to cope with a high enemy launch rate. If necessary, the F-15 could undertake covert attacks as well.

The Vought Miniature Vehicle was designed in competition with a General Dynamics model which used a sensor mounted on gimbals and an external motor array for lateral control. Once Vought won the contract, extensive use was made of simulators during development, the goal being to reach as high a degree of confidence as possible. This involved clean rooms, vacuum chambers and a drop-test facility. In the latter, the Miniature Vehicle tracked an infrared source during a 2-second free-fall. The inevitable development problems cropped up. The infrared sensor had a sensitivity problem and limited attack geometry. Finding a solution delayed the program only slightly. The computer was given a larger capability and the vehicle structure was strengthened after vibration testing.

By late January 1982, a mock-up of the ASAT weapon had been fitted to a modified F-15. By May 1982, the F-15, ASAT and ground facilities were being merged into a functioning unit.

The target satellite built by Avco for the ASAT tests was spherical and battery-powered. Ingeniously, the Avco target can change its thermal signature to match that of Soviet satellites by using a system of louvers.

Recent Soviet anti-satellite developments In early 1980 with US-Soviet relations at their lowest ebb in many years, the time for restraint was over. On April 3, 1980 Cosmos 1171 was launched into a 626 by 600-mile (1008 by 964-km) orbit by a C-1 from Plesetsk. Its characteristics instantly labelled it as an ASAT target. The interceptor Cosmos 1174 was launched by an F-1-m on April 18 into an initial orbit of 252 by 77 miles (405 by 124 km). Almost immediately, it fired the on-board engine and went into a 637 by 225-mile (1025 by 362-km) attack orbit.[67] The interception came during Cosmos 1174's second orbit over Leningrad.

The test was an inauspicious beginning to the renewed Soviet ASAT program for the Cosmos 1174 interceptor missed. One source gives the miss distance as 4.9 miles (8 km), another as 37 miles (60 km). This was apparently not the end of the story, however. A little less than six hours after launch, Cosmos 1174 maneuvered again, this time going into a 1032 by 236-mile (1660 by 380-km) orbit to make a second pass of the Cosmos 1171 target, off the east coast of the United States. The closest approach was about 43 miles (70 km). The next day Cosmos 1174 made a small maneuver that resulted in a third flyby. Taking place over the eastern Pacific Ocean in the area of Easter Island, the separation distance on this occasion was less than 12.4 miles (20 km). Finally, on April 20, the Cosmos 1174 interceptor blew itself into fragments.

What was the purpose of Cosmos 1174's two-day flight? One possibility is that it tested an advanced interceptor able to make a second or even a third attack. Alternatively, it may have been an inspector designed to take multiple looks at a target. In any event, it gave a remarkable performance. There were suspicions into the bargain that both Cosmos 1009 and 1174 used a new infrared guidance system instead of radar.

The Soviets' next test occurred nine months later. This time the profile was straightforward. The Cosmos 1241 target was launched on January 21, 1981 into a 627 by 605-mile (1009 by 973-km) orbit, nearly identical to that of the previous target. Twelve days later the Cosmos 1243 interceptor was sent up into a 629 by 183-mile (1013 by 294-km) orbit, catching up with the target and passing within lethal range after 2.5 orbits.[68] Once again interception occurred over Leningrad. A re-entry burn then sent Cosmos 1243 plunging to destruction in the atmosphere.[69,70]

In the penultimate test six weeks later, the Cosmos 1258 interceptor was launched into a 634 by 186-mile (1021 by 299-km) initial orbit (which resembled that of its predecessor) to fly past the target 2.5 orbits later, just as in the previous test. The Cosmos 1258 interceptor then exploded.[71] This interceptor is believed to have used radar. On September 23, 1981 Cosmos 1310 — a diagnostic and calibration satellite — was placed in orbit, bringing the year's ASAT activity to a close.[72]

In the public press, there were reports that the Soviets were developing two new ASAT systems. The first was a prototype manned battle station. *Aviation Week & Space Technology* carried a series of reports in its Washington

Round-Up column that the Salyut 6-Cosmos 1267 space station was equipped with multiple infrared guidance homing interceptors. They were carried aboard the Cosmos 1267 space station module[73]. which had been launched by a D-1 booster on April 25, 1981. During the next two months, it maneuvered extensively and released a capsule. Cosmos 1267 finally docked with Salyut 6 on June 19 at a time when Salyut 6 was unmanned, the last crew having left the previous month.

The first spacecraft of this type, Cosmos 929, was flown independently for six months in 1977-1978. It, too, released a capsule after one month and maneuvered but did not dock.[74] Salyut 6-Cosmos 1267 orbited the Earth until July 29, 1982 when it was de-orbited and re-entered over the Pacific.[75]

Surprisingly, the initial interpretation of the 'battle station' was that it was an ASAT. The Soviets were expected to add more spacecraft with interceptor missiles to give it a capability against ballistic missiles.[76] A year later, the interpretation changed. Salyut 6-Cosmos 1267 was now described in the Western press as a DSAT (defensive satellite). The homing vehicles were a space station defense system to prevent any US attempts to interfere with Soviet satellites.[77] This was similar to the early speculation about the Soviet ASAT tests. Others believed Cosmos 1267 was an innocent space station module and that the ASAT speculations were wrong.

The other reported ASAT corrected a serious deficiency in the Soviet program. It was a direct-ascent system to attack geosynchronous satellites with a nuclear warhead as the 'kill mechanism'. The lethal radius would be large enough to allow for guidance inaccuracies. It had long been realized that the Soviets had this capability and because it depended on the technology already used to put satellites into geostationary orbit there was no need for the Soviets to test the system.[78] It meant that communications, ELINT and early warning satellites could now be destroyed. Their destruction would allow the Soviets to make a clean sweep of Western space systems.

The Soviets were meanwhile active in the propaganda arena. US representatives at the ASAT talks had considered Soviet hostility to the Shuttle a negotiating ploy to be dropped after gaining US concessions. Contrary to expectations, after the collapse of the talks the Soviets intensified their campaign against the Shuttle. They claimed, for example, that the first flight's goals were primarily military and that the crew had tested a laser-weapon sighting device. The Soviets claimed as well that two-thirds of the Shuttle flights would be military. Actually, only one-third would be (a fact the US has never concealed). Moreover, the military background of Robert Crippens, Columbia's co-pilot, was viewed as sinister even though most astronauts and cosmonauts are military pilots.[79]

In fact, the Shuttle represents a capability the Soviets will not be able to match until the 1990s. It can carry large payloads into space on a routine basis and is re-usable, opening up the space frontier much as the railroad opened up the American West. It makes the Soviet space program, still tied to expendable boosters and ballistic re-entry capsules, look old-fashioned.

Soviet criticisms of the Shuttle must be seen in their political context. The Soviets present a facade of complete innocence despite their extensive space programs with military aims. At a cosmonaut reception not long after Columbia's first flight, Soviet leader Leonid Brezhnev intoned 'may the shoreless cosmic ocean be pure and free of weapons of any kind. We stand for joint efforts to reach a great and humanitarian aim — to preclude the militarization of outer space'.[80] It is a situation full of ironies.

The seven-hour nuclear war

About a year later, in June 1982, the UN held a meeting on disarmament. The Soviets again stressed the control of space weapons. On June 6, shortly before the UN convened, the Soviets launched Cosmos 1375. Its orbit of 634 by 615 miles (1021 by 990 km) and 65.9° inclination marked it out as an ASAT target. Launching of the Cosmos 1379 interceptor came 12 days later, like the first 1981 test. This, and the rather regular scheduling, indicated that missions of this kind were routine training exercises similar to Program 437's CELs. Cosmos 1379 entered a low orbit from Tyuratam and 1 hr 16 min later, maneuvered into an orbit of 628 by 607 miles (1010 by 977 km). The flyby of the target came 3 hr 10 min after launch as the Cosmos 1379 interceptor neared apogee. After passing the Cosmos 1375 target, the interceptor continued in orbit for another 30 min before re-entering.

The ASAT test was part of a wide-ranging Soviet military exercise. During a seven-hour period on June 18, the Soviets fired two SS-11 ICBMs from their operational silos following quickly with the launch of an SS-20 IRBM. The Soviets also fired a submarine-launched ballistic missile from a Delta class submarine in the White Sea. To complete the test, two ABM-X-3 missiles were fired against ICBM re-entry vehicles. The tests were linked by the Soviet command and control network.

For those seven hours, the Soviets fought an imaginary nuclear war — a first strike against the US and Europe and a follow-up second strike by submarine missiles, then the interception of the retaliatory US counterstrike. The exercise indicated that ASATs were a key part of Soviet nuclear strategy. In a real attack, they would destroy US reconnaissance satellites. This would prevent the US from determining which Soviet ICBMs had been fired, or locating such targets as ballistic-missile submarines or silo re-loading.[81] President

Ronald Reagan gave an address on space policy at Edwards AFB a month after the Soviet ASAT and war-fighting exercise, stressing the need for the US to deploy ASAT weapons. This was a clear response to the Soviet test.[82] By October 1982, an F-15 anti-satellite combined test force had been established to carry out the ASAT launches.[83]

Test firings of the ASAT missile have suffered from continual delays. At the start of 1982, they were intended to take place before the end of the year. In early 1983, a new date of late summer was set. The F-15 ASAT, according to one source, will not try to intercept satellites at once, the first firing testing only the two-stage missile. It will be aimed at a point in space and the Miniature Vehicle will not be carried. Once the basic missile has proved itself, the second launching, set for the autumn, will test the Miniature Vehicle's infrared sensor. The actual interception of a target satellite is scheduled for the third flight.

As both nations prepare advanced anti-satellites like these — systems which by 1983 were nearing completion — the next generation of weapons can be dimly discerned on the distant horizon, and it is these weapons of tomorrow that we consider next.

Looking like an Egyptian temple, the Cobra Dane radar rises above the bleak shore of Shemya Island in the Aleutians. This tracking station was the only one to pick up several of the Soviet ASAT tests before they re-entered. Only 2 miles wide by 4 miles long (3.2 by 6.4 km), Shemya Island is windswept, overcast and covered with fog for most of the year. Temperatures range from − 6.7 to 10°C (20 to 50°F).

The F-15 Anti-satellite

A new departure in ASAT technology, the F-15 launched system is an intricate and expensive one which has several unusual features. It can be flown almost anywhere to accomplish an interception, obeying commands from the Space Defense Operations Center at the heart of Cheyenne Mountain.

It is launched from an F-15 Eagle in subsonic, straight or level flight for most attack profiles. After the two-stage missile has burned out, the Mini-Vehicle which it carries comes into play. This sophisticated homing unit destroys by direct collision with the target. It hits the luckless satellite with the impact of a shell from a battleship's main gun.

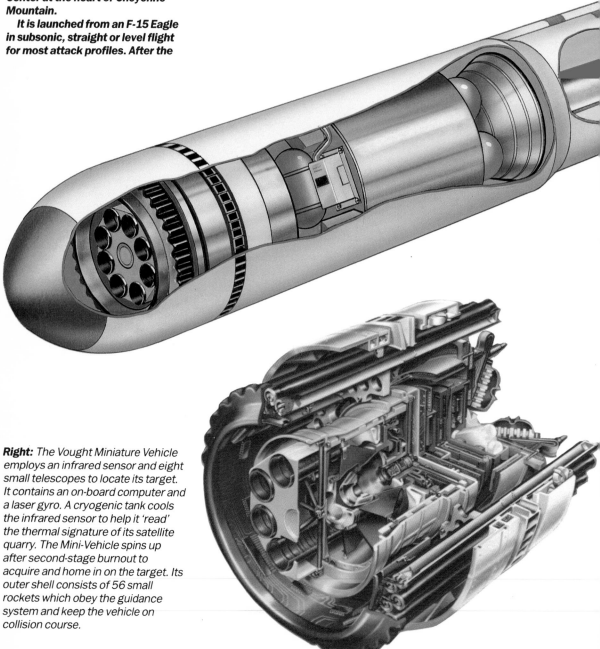

Right: The Vought Miniature Vehicle employs an infrared sensor and eight small telescopes to locate its target. It contains an on-board computer and a laser gyro. A cryogenic tank cools the infrared sensor to help it 'read' the thermal signature of its satellite quarry. The Mini-Vehicle spins up after second-stage burnout to acquire and home in on the target. Its outer shell consists of 56 small rockets which obey the guidance system and keep the vehicle on collision course.

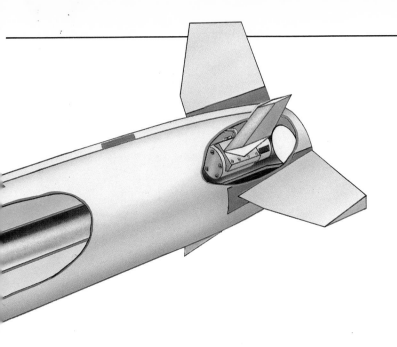

Left and **below:** Based on the Boeing Short Range Attack Missile, the F-15 ASAT has two small fixed fins and three large movable fins for control during ascent. For satellites in higher orbits, the F-15 fires the missile during a supersonic climb, adding speed to the ASAT. The automatic launch asks little more of the pilot than a video game, thanks to the head up display. The principal advantage of the F-15 ASAT is its mobility. It makes a very difficult target, whereas a fixed-based ASAT — like that on Johnston Island — is by comparison a 'sitting duck' for enemy forces.

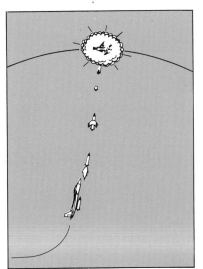

Left: The booster which propels the Vought Mini-Vehicle uses solid-fuel rockets. Unlike Program 437 and the Cosmos ASATs, the F-15 version is mobile. The concept of such a 'satellite killer' was first mooted 25 years ago. Only in the late 1970s, with the revival of the Soviet ASAT program, was it actually developed. Here the ASAT is loaded onto an F-15 at the Boeing Development Center.

Chapter 5

LASERS

'Progress (…), without a parallel moral progress, which is continuous and internal, (…) develops what is lowest and cruellest in man, making him a machine possessed by machines, a number manipulated by numbers; he becomes what Papini called "a raving savage, who, to satisfy his predatory, destructive and licentious instincts, no longer uses a club, but has the immense forces of nature and mechanical invention to draw upon".'

Albino Luciani,
Illustrissimi (1978)

Looming large in the plans of both the US and the USSR are energy beam weapons. Lasers and particle beams have the potential to change warfare much as the atomic bomb did 40 years ago. They could become an impenetrable defense making the long-range missile obsolete — and with it the nuclear doctrine of Mutual Assured Destruction. Alternatively, they would simply be a new way for man to kill his brother. Whatever their applications, these weapons are real and their potential must be faced.

Speculation about 'death rays' began before the first laser light gleamed. H G Wells started it all in 1898 in his classic novel *The War of the Worlds* which depicted Martian invaders ravaging the English countryside by projecting narrow beams of intense heat against their targets. Metal softened and melted, glass shattered and water turned instantly into steam.[1] No longer confined to the realms of the imagination, beam weapons are today fast-approaching operational readiness.

Not long after *The War of the Worlds* was published, scientific discoveries were made that would ultimately turn Wells' 'heat ray' fiction into fact. The period between 1900 and the 1930s was a golden age of physics. Max Planck, Niels Bohr, Max Born, Werner Heisenberg, Ernest Rutherford,

A target drone is destroyed in mid-flight by an experimental laser weapon. Kirtland Air Force Base, New Mexico.

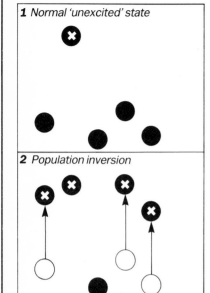

1 *Normal 'unexcited' state*

2 *Population inversion*

3 *Photons released*

4 *Photons strike 'excited' electrons*

How a laser works. Most electrons ordinarily follow a close orbit about the atomic nucleus (1). When the lasing material is pumped, they become 'excited' (2). When sufficiently 'excited', the electron emits a photon (3) which begins a dynamic chain reaction (4), before the electron resumes its lower orbit.

James Franck and many others were laying the foundation for the science of quantum mechanics.[2] This branch of physics deals with the structure of the atom as well as the nature of light. It serves as the basis of the transistor and microcircuit as well as the laser.

The term laser means Light Amplification by Stimulated Emission of Radiation. To see just what this jaw-breaking title means, we must retreat to the atomic level. The atom can be described as resembling a miniature solar system. At its center is the heavy nucleus which plays the part of the Sun. Orbiting it are the electrons. Normally these electrons follow a close orbit about the nucleus. If the electron is hit by a photon of light of the correct wavelength and energy, however, the electron will absorb it and jump to a higher orbit or energy level. The electron is then said to be 'excited'.

For the laser to work, most of the electrons in the lasing material must be raised to this higher energy level. To do this, the lasing material is 'pumped' by either an external light source or a strong electrical current. The goal is to pump the lasing material until most of the electrons have been raised to this higher energy level, a process called population inversion.

This higher energy level is not permanently stable and is only maintained for a brief period. The electron emits a photon of a specific narrow wavelength and now, rid of the extra energy, drops back into the lower orbit. This emitted photon then encounters another excited electron. Its presence stimulates the electron to emit its photon, so there are now two photons with an identical wavelength and phase. The next encounter results in four identical photons, then eight, 16, 32 and so on. It is an optical chain reaction reaching awesome proportions.

At either end of the lasing material are mirrors. One is fully coated with reflecting material; the other is lightly coated so that most of the beam can shine through. When the photons strike a mirror, they are reflected back into the lasing material where they can stimulate the emission of still more photons, increasing the power.[3]

Lasers have two operating modes: *pulse* in which all the energy is concentrated in a single burst, or *continuous-wave* in which the beam is sustained. The narrow intense beam of a laser has little tendency to spread. This is because the light waves are in phase or 'coherent'. Think of the coherent light of a laser as a line of marching soldiers all in step and all going in the same direction. The soldiers are the light waves in the beam. Incoherent light, produced by a normal light bulb, can be likened to a crowd milling about aimlessly. The laser's narrow beam allows the full power to be concentrated on a very small area. The light from a laser also has a single distinctive frequency or wavelength. There is very little spread from this (a laser beam may be only 1.86 miles [3 km] wide after travelling 250 000 miles [400 000 km] to the

Moon).[4] This is a consequence of the laser process itself, and per specific frequency depends on the lasing material. A light bulb, on the other hand, gives off light across a wide band of frequencies.

Two types of laser — gas and chemical The theoretical background of the laser can be said to have existed since the 1930s. In the late 1950s theory, understanding, technology and interest came together. First proposed by A L Schawlow and C H Townes in 1958, it was two years before the first workable device was built in the form of a 'pulsed laser', by Dr Ted Maiman of the Hughes Aircraft Company. A ruby rod was used as the lasing material, the light source being a Xenon discharge tube similar to an electronic flash gun.

At first the invention was called 'a solution in search of a problem', but it was not long before applications began to appear. Because the light is of a single frequency, it can be used in precision tests of optical equipment. Land surveyors use lasers to measure distances and directions with great accuracy. One early use was in delicate eye surgery. The uses have continued to multiply like the emitted photons themselves.

Perhaps because the laser so closely resembles the 'death ray' of science fiction, military applications generated high interest. Even the early low-power lasers could burn through a razor blade at close range. Although the beam has little power, all of it is concentrated into a spot only a few wavelengths of light wide (75 watts falling on an area one wavelength to a side is equivalent to 300 000 million watts on one square inch).[5] The 'death ray' seemed at hand.

In real military terms it was a parlor trick. The range was only a few inches, the amount of damage trivial; a pinhole at best. To be militarily useful, the power would have to be increased at least a million and perhaps a billion times. With the ruby rod laser, this was not possible. Only a small percentage of the energy put into the rod was converted into laser light. Most of the energy became heat, and even at modest power levels this heat would destroy the rod.[6]

The breakthrough came in the form of the gas laser. The first gas laser was very simple; a glass tube containing gas, much like a neon sign. In the first example, the gas was a mixture of neon and helium, so formulated as to support the laser reaction. The energy, provided by electrodes, causes an electrical discharge which pumps the gas. At each end are mirrors, but unlike the ruby version this laser allows a continuous beam. The electrodes continuously elevate the electrons to the higher energy level and even though some fall back, enough excited electrons remain to sustain the lasing action. Because of the longer tube length compared to the ruby rod and the continuous action, the beam is more coherent. Rather than hitting it with brief pulses, the continuous beam can dwell on a specific area of the target, heating it up.

Subsequent research indicated that almost anything can be made to lase. Other gases for instance include argon, carbon dioxide and krypton. An ordinary tube laser still has heating problems; as more and more power is pumped into it, the gas heats up and the laser becomes less efficient.

The concept evolved into the gas-dynamic laser and this development turned a laboratory curiosity into a weapon. In the mid-1960s, the best gas laser projected a beam with a power of 100 watts. Then, in 1968, at the Avco Everett Research Laboratory, a carbon dioxide laser was developed that produced a 60 000 watt (60 KW) continuous-wave beam. Higher power levels were attainable (although with a loss of coherence).[7] Inch-thick steel is swiftly pierced by 10 000 watts at close range.

The gas-dynamic laser uses a flow of gas rather than a sealed tube, the lasing action taking place in a chamber called the optical cavity. As in the gas-tube laser, there are mirrors at opposite ends. The gas is pumped through the optical cavity at a pressure below that of the atmosphere. The first gas to be used was carbon dioxide (though carbon monoxide can also be made to lase). The gas flows at supersonic speeds between the mirrors and overall construction resembles that of a wind tunnel.

Powerful electrodes cause the lasing action and millions of watts of power pass through the electrons to bring about the population inversion, most of which ends up as heat. Too much heat can cause chemical changes in the gas that could stop the lasing action. The carbon dioxide, for example, may break up into carbon and oxygen and nitrogen can combine to form oxides of nitrogen. Despite such problems, however, the gas dynamic laser operates at high power. The flow of gas carries the heat off before it can melt the optical cavity or contaminates can poison the reaction. The spent gas is either exhausted to the outside or run through a heat-exchanger to be cooled and then fed back into the optical cavity.

The gas-dynamic laser has been further developed into the chemical laser. One disadvantage of the gas-dynamic laser is the huge amount of power it requires. The chemical laser, on the other hand, does not require an external power-source at all. The energy to pump the lasing gas is provided by chemical reaction. Again, carbon dioxide was first used in a chemical laser and has proven most versatile, even being made to lase with sunlight.

In a chemical laser, carbon monoxide is burned with nitrous oxide. The reaction produces carbon dioxide with its electrons at the higher energy level. The energetic state is maintained by expansion through a bank of supersonic nozzles which establish the necessary flow conditions for the lasing

action. Mirrors at either end reflect the emitted photons. The hot gas is then exhausted to the outside and in this respect the system resembles a rocket engine.[8]

Though the design of each of these lasers is different, the fundamental physical principle is the same in each case. Later designs use other gases. Fluorine has found common usage in various mixtures including deuterium fluoride, hydrogen fluoride and krypton fluoride. The use of fluorine however means that the exhaust is extremely poisonous. What is more, hydrofluoric acid will soak into skin and attack the calcium in bones; it will even dissolve rocks. The only containers able to hold it are made of wax and Teflon.

The laser field is obviously a dynamic one with many unknowns. Lasers are not so much designed as discovered. A new additive or a new mirror design, for example, can radically change efficiencies. The physics is still not completely understood and it must be noted that there is still a large measure of speculation and experiment alongside careful analysis.

A laser beam burns through a Plexiglas sheet. Different materials experience shock effects of different magnitudes under laser attack. Plexiglas experiences 100 times the shock effect generated in metals at the same light level.

The laser's military applications

What makes the laser an effective weapon is the fact that it strikes at the speed of light — more than 186 000 miles per second (299 274 m/sec). There is virtually no need to anticipate the future position of a target; a laser beam can cover 100 miles (161 km) in 0.000538 seconds. In that time, an orbiting satellite moves just 13.8 ft (4.21 m). Also, a laser gives no warning of an attack; there is only the blinding flash as it strikes. This is in sharp contrast to a direct-ascent ASAT, which requires several minutes to reach the target satellite and so allows time for the target to take action (either to maneuver, release decoys or warn that an attack is underway).

A laser's optical system is also comparatively light-weight. It can be turned quickly, moving from one target to the next in short order, and able to cope with targets approaching from different directions. A common optical system allows both ranging and firing and might include a television camera, laser rangefinder and tracker.[9]

When the energy of a laser beam strikes a target, it is converted into heat by a process called thermal coupling which depends on both the wavelength of the laser and the target material. Thermal coupling increases at shorter wavelengths. Tests with low-powered lasers indicate that aluminum absorbs about three to seven per cent of the laser light, the exact amount depending on the alloy. Another factor is the material's surface condition. A smooth, shiny surface will absorb less energy than a dull surface. At higher energy levels, the thermal coupling increases significantly. This heating can weaken a target's structure and if it is under stress may break it up.

With enough energy, the material starts to vaporize, giving off a metal-gas cloud which absorbs some of the laser's energy. If this cloud is sufficiently dense, it may actually stop the beam reaching the target. Not until the cloud disperses can the laser resume the attack. There are two solutions to this. One is to pulse the laser, thus allowing the cloud to disperse. The other is to scan across the target rather than concentrating on one point.

The sudden vaporization of the material under attack causes a shock wave in the target almost like an explosion. In some materials, such as plastics, this may mean shattering. At lower power levels, the degree of shock effect is directly related to the type of material. Plexiglas, for example, experiences 100 times the shock effect generated in metal at the same light level. In graphite and ceramic materials, the effect is ten times greater than in metals.

Above certain power levels, the shock damage no longer depends on the material; there is a potentially more lethal effect. The vaporized metal can generate X-ray radiation, damaging the crystal structure of the metal and electronic components.[10] Whereas metal heated by a flame will glow red, giving off visible light, the material vaporized by the laser will 'glow' in X-rays.

What this means, in practical terms, was shown in a ground test. An F-104 was illuminated with an experimental carbon dioxide gas-dynamic laser. The windscreen became incandescent and everything in the cockpit burned up.[11]

Lasers have been fired against materials ranging from sheets of titanium to containers representing ICBMs. Some of these tests were against airborne targets. A carbon dioxide laser was fitted with a telescope system at the Sandia Optical

Range at Kirtland AFB, New Mexico, and fired against a propeller-driven target drone. The beam struck the fuselage; after a brilliant flash, the forward fuselage separated and the wing fell off.[12]

The next to destroy a US target was the Army's Mobile Test Unit — this had an Avco gas-dynamic carbon dioxide laser mounted aboard a modified LVTP-7 amphibious-assault vehicle. This particular laser is said to have been based on a 10 to 15 KW continuous-wave metal-cutting laser. A turbine-driven generator provided electrical power for the laser and two large radiators mounted on the LVTP-7's sides dissipated the heat. The MTU was built to test the ability of the laser and its sighting system to endure travel over dirt roads and rough terrain.[13] After the road tests were completed at Redstone Arsenal in Alabama, the MTU shot down both aircraft and helicopter drones in 1976.

In late 1977 and early 1978, a Navy-built deuterium fluoride chemical laser destroyed four TOW wire-guided anti-tank missiles. This was a significant achievement since they were pulling several hundred G's when they were hit. It was the first time a laser had destroyed such a fast-moving target. The test was primarily to examine the tracking system's ability to hold the beam on a small, maneuvering target. It had to detect and follow the small missile without the use of radar. The Navy laser was also fired against tethered Huey helicopters at the Navy's San Juan Capistrano, California test site.[14] Built to pave the way for the Navy's 2.2 megawatt Sea

US Navy laser test site at San Juan, Capistrano, California. Here the pioneering high-energy laser program tested a prototype US laser weapon.

Lite laser (the first with enough power to be considered an operational weapon), the laser has an output of 400 KW.

Problems with laser weapons

The laser's worst enemy is the Earth's atmosphere. Out in the real world, things are not so tidy as in an air-conditioned, dust-free laboratory. Considerable effort has been devoted to examining atmospheric problems. A particular concern is thermal blooming. The air is like a weak lens; it can bend light. The amount of bending is called the index of refraction. When a laser beam passes through the air, the air is heated up. This changes the index of refraction, smearing the beam and throwing it out of focus. As much as 90 per cent of the beam's energy can be lost. One remedy is to pulse the laser. With proper timing, the air does not heat up enough to de-focus the beam. Air currents, or the motion of the laser as it tracks the target, insure that the beam always passes through cool air.

The procedure does have limitations. If the pulse is too short, or the interval too long, not enough energy will hit the target. The other extreme is just as bad. Too much energy in each pulse will cause an electrical breakdown of the air. The atoms absorb so much energy that their electrons are stripped off, with a remarkable effect. The ionized gas appears as bright blobs tracing out the path of the laser beam.

Although most are small, they can grow to perhaps an inch in size,[15] and the energy that causes them to form is wasted. They also shield the target from the beam. The amount of loss is dependent on the wavelength of the laser and pulse duration. As a general rule, the shorter the wavelength, the fewer problems with ionization.

There are other problems. The air is in constant motion, so that wind and turbulence take their toll. This is because the index of refraction is altered by differences in air density. As the beam passes through the air, it can hit areas of strong refraction, followed by areas of weak refraction. This spreads the beam. Density, of course, also changes with altitude. In addition, the atmosphere is not composed of oxygen and nitrogen alone. It contains carbon dioxide, water vapor, haze, fog, salt spray, smoke and other aerosols which can both scatter and absorb the laser light. Particles in the air can cause electrical breakdown at much lower power levels than normal. An example of the effect of atmospheric factors is given by a United Technologies' study which compared different types of 25 KW continuous-wave lasers. In each case, the beam was 20 in (51 cm) wide with the target at a distance of 1.9 miles (3 km). A carbon dioxide laser hit the target with only 100 watts/sq cm; which can be compared to the thermal output of a soldering iron. A deuterium fluoride laser with the same power, beam diameter and range delivered 4 KW/sq cm, an intensity 40 times greater.

The designer must perform a balancing act, trading off laser frequencies, pulse energy, duration and rate against such factors as the environment in which it will operate, the nature of the target and the range and the design of the optical system.[16] Although a carbon dioxide laser is not affected as much by turbulence it suffers from absorption by water vapor. It could however be useful as a defense system for bombers. At high altitude, there is little water vapor and interference from the wake turbulence of the aircraft is avoided. In contrast, the primary concern about shipboard use is the laser's ability to penetrate humid sea air. Here the deuterium fluoride laser has distinct virtues.

In summary, it can be stated that atmospheric effects have, so far, limited development of battlefield lasers for anti-aircraft use. One estimate ranks laser anti-aircraft weapons as possessing ranges comparable to guns, but without the range of missiles. Nor are lasers all-weather weapons, unlike radar-directed guns and missiles which operate equally well in clear air or dense fog.

Anti-satellite laser developments   Lasers have been studied for ASAT, DSAT and ABM roles. A variety of methods have been proposed: ground-based systems; a laser aboard an aircraft; and lasers operating from space. An operational ground-based ASAT laser is feasible in the very near future. In the mid-1970s, the Air Force actually planned to build two high-power lasers as anti-satellite weapons, one on a mountain near Kirtland AFB, the other at China Lake, California. The project was cancelled due to budget cuts.[17]

A subsequent proposal was to develop the 2.2 megawatt Sea Lite laser that the Navy planned to build at White Sands. With a new pointing-tracking system, this would demonstrate the ability of a ground-based laser to destroy low-orbit satellites. Tests were anticipated as early as 1986 after the Navy had completed its work.[18]

A ground-based ASAT laser involved compromise between atmospheric losses and ease of assembly. It faces a different atmospheric problem than an anti-aircraft weapon. A target for an anti-aircraft laser flies at low altitude: a cruise missile skimming the waves, for example. The attacking laser has to fire at a low elevation through a thick layer of atmosphere. An ASAT laser, on the other hand, would be firing at a high elevation, its beam passing through a minimal layer of the atmosphere. Consider the analogy of a smoggy day. On the horizon, a band of brown haze is seen, yet directly overhead the sky seems clear. The pollutants are in fact evenly distributed. Toward the horizon, one is looking through a greater thickness of pollutants. Siting of the laser is therefore important. A mountain top site can get the laser above most of the atmosphere. Ideally, the area should have a minimum of atmospheric turbulence and year-round cloud-free weather.

Counter-balancing the problems of the atmosphere is the comparative ease with which a laser can be constructed and operated. A ground ASAT laser avoids the packaging problems of an airborne or space laser. It could take up as much space as necessary, drawing its power from the national grid. The Hoover Dam (or the Bratsk high dam) could even be used. The lasing gas could be kept in large tanks, re-supply needing nothing more than converted tanker trucks. Any malfunction could be quickly repaired by on-base maintenance personnel.

A major Soviet laser-development center is located at Krasnaya Pahkra, about 9 miles (15 km) south of Moscow's Vnukovo airport. US Intelligence sources believe that tests have taken place there of an anti-satellite carbon dioxide laser pumped by an electron-beam system. The output is estimated to be several hundred thousand watts, powerful enough to disable US Big Bird and KH-11 reconnaissance satellites. The Soviets are also reported to be working on a more powerful multi-shot laser with the object of disabling satellites at altitudes of about 3000 miles (5000 km).[19,20]

A more advanced US ground-based laser might use an orbiting mirror to reflect the beam to the target. The laser could be in a super-hardened emplacement able to withstand a nearby nuclear explosion. With accurate ICBM guidance systems, the laser would need such protection from incoming fire. Unlike an ICBM silo, however, the laser would be self-defending.

One approach for the relay mirror could be for it to be placed aboard missiles on the ground. The boosters would be kept at instant readiness. By keeping the mirrors on the ground, they would not be vulnerable to a Soviet ASAT strike. When a Soviet attack was detected, they would be launched. Once in orbit, the mirrors, which reports suggest could be up to 100 ft (30 m) in diameter, would deploy. The ground laser would fire at the mirror which would refocus the beam energy onto attacking ICBMs. If the risk of a Soviet attack on the mirrors was deemed sufficiently minimal, they could be permanently stationed in orbit. Needless to add, building space mirrors of this capability presents formidable problems.

If technical difficulties can be overcome the ground laser/space mirror could be an interim step between the limited ground-based laser and the full-scale orbital ABM. The heavy and technically demanding parts (laser gun, fuel, power sources, etc) are left on the ground. Using an orbital mirror to replay the beams means the range can be extended. A normal ground laser can only cover the terminal phase of an attack. A space mirror could cover the mid-course of the missile's trajectory or even the earlier boost phase.

A ground laser/space mirror would still have the problems of the earth's atmosphere. There is the problem as well of coordinating the laser and the mirror, as they are separated by perhaps thousands of miles. This involves ensuring that the mirrors are positioned so the incoming beam can strike the mirror and then be reflected accurately to the ICBM. The mirrors would also require aiming systems to detect the ICBM and to point the beam onto the target.

A satellite's weakness Non-maneuvering satellites are almost 'sitting ducks' for a laser marksman. Depending on the geometry of the orbit, a satellite can be above the horizon for several minutes. In theory, even if a large percentage of the beam's energy is lost to the atmosphere, the laser could be kept firing long enough to cause damage.

A satellite is vulnerable to laser attack in a number of areas. Most satellites use infrared horizon scanners, as well as Sun and star seekers to orientate themselves in space. A laser could burn these out, leaving the satellite to tumble helplessly in space. Even a comparatively low-power laser could do this.

Electronic and other components are vulnerable to overheating. When sufficiently heated, a transistor begins to draw more current and produces even more heat — a rising spiral which ends when the transistor fails. A laser could also overload a satellite's thermal-control system. Typically, satellites are equipped with louver-like radiators; under laser bombardment, they would pump heat into the electronics instead of carrying it away, with disastrous results. Optical systems, such as the cameras of a reconnaissance satellite, are particularly vulnerable. The coatings on lenses and mirrors are easily damaged. The glue that holds achromatic lens elements together fails under laser light. Glass itself is damaged by concentrated light of tens to hundreds of megawatts per square centimeter.

To maintain temperature control, US satellites use polished-aluminum surfaces and mylar insulation. The latter is thin plastic with a coating of gold. This layer may be only a few atoms thick. These materials would be able to take some laser heating, but for how long is open to question. Soviet satellites are of a more rugged construction which, ironically, may make them even more vulnerable.[21,22]

Another vulnerable part of the satellite is its solar panels. A satellite might survive with some types of damage but if deprived of power it is dead. The solar cells' electrical connections can be melted. Heat can render the cell covers opaque and weaken the glue holding the cells together. Both gallium arsenide and silicone solar cells can be destroyed at temperatures under 600°C.[23]

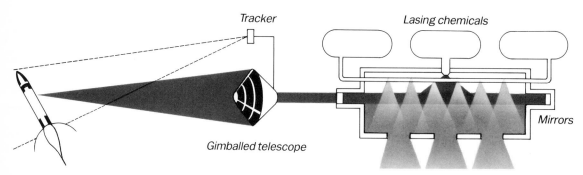

Inside the laser turret

Tracker

Lasing chemicals

Gimballed telescope

Mirrors

Exhaust gas

Top: First US air-to-air tests of a high-energy chemical laser were made from this converted Boeing 707 called the Airborne Laser Laboratory (ALL). The diagram (above) shows the mechanics of the chemical laser and pointing system installed in the ALL. The laser turret is mounted on the 'hump' behind the crew compartment and the fuselage is filled with laser fuel and equipment.

Right: The inside of the NKC-135 ALL, looking towards the aircraft's tail. The cylindrical laser combustion chambers are in the foreground. In the rear are the laser fuel tanks. The various equipment necessary to operate and monitor the laser takes up most of the ALL's internal volume.

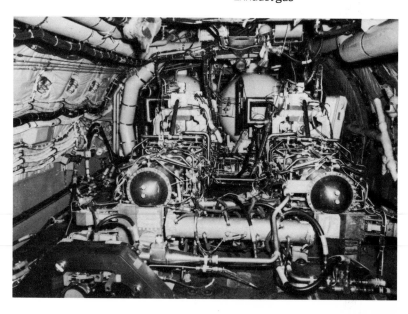

Airborne lasers — advantages and problems

The most serious limitation inhibiting a ground-based ASAT laser, as noted earlier, is the atmosphere. A certain amount of energy will inevitably be lost. Even at a desert site, clouds and rain will interfere for a number of days each year.

An alternative is to put the laser aboard an aircraft. There are a number of advantages to this. The aircraft laser can go anywhere in the world. It could also fly parallel to the satellite's ground track. With a 500 mph (804 km/h) ground speed, an aircraft could chase the satellite, extending the attack time. The aircraft could climb above the weather; above 30 000 ft (9144 m), most of the absorption factors that occur at low altitude no longer apply. Not only is the air considerably thinner, there is little if any water vapor. A carbon dioxide laser could therefore be used,[24] which eliminates the need to handle the extremely toxic fluorine gas.

There are, however, major technical challenges in building an airborne laser. The chemical laser's need for a large volume of gas means limited firing time. An aircraft, even a large transport, is limited in volume and weight. Only so much lasing gas can be carried (the exact amount is uncertain). One estimate gives a total firing time of 20 to 30 seconds; another extends this to a full minute.[25] Gas dynamic lasers, on the other hand, can recycle their gas but require several megawatts of power, about ten times the available power of a jet transport or strategic bomber. A fuel cell to produce such power was estimated to weigh 2500 lb (1134 kg). A battery system would weigh 3000 lb (1360 kg); another 1000 lb (454 kg) would be needed for the transformers and converters. For the above reasons, the most likely candidate for an airborne laser is a chemical system — despite the fuel problems. To compensate for this, the aircraft could land and refill the tanks.

The operating environment of an aircraft in flight also poses questions. When consideration was first given to an aircraft test laser in the early 1970s, there were questions as to whether the air flow around the aircraft would distort the beam. Could a laser be aimed from a moving, vibrating platform? Finally, it was not clear if the laser's optical system could be kept in alignment because of the flexing of the aircraft.[26]

To answer these questions, the Air Force modified a 707 tanker to carry a laser in a turret atop its fuselage. The NKC-135 Airborne Laser Laboratory (ALL) was first fitted with a low-power carbon dioxide laser to investigate basic problems using another KC-135 as the 'target'. The latter had equipment to measure the intensity and dispersion of the ALL's laser. In general these tests showed that the unknown areas were possibly not as formidable as had been feared.[27]

The NKC-135 ALL was subsequently equipped with a 400 KW carbon dioxide chemical laser built by Pratt & Whitney in which spent lasing gas is vented through defusion doors on the aircraft's underside. The interior of the NKC-135 is taken up with the split stage combustion chamber and fuel tanks. As in the case of the other test lasers, ALL was meant to evaluate the limitations and advantages of laser weapons and to better understand their physics.[28,29]

The ALL attempted to destroy a Sidewinder air-to-air missile fired from an A-7 over China Lake, California on June 1, 1981. It missed. A second attempt two days later also failed. After this, the Air Force refused to comment.[30] It was rumored later that the reason for the failures was unanticipated vibration. Despite ALL's less than auspicious public debut, it represented a major technological achievement. A laser and its aiming system had been put into an aircraft and operated in flight. Some hold that this is only one step away from putting a laser into space, albeit a large one.

The near-perfect vacuum of space could well be considered the laser's natural home. More energy can be deposited on the target rather than scattered. Other factors are also in its favor. Most designs of gas lasers require low optical-cavity pressures; because of the vacuum, the gases can be vented overboard. Space eliminates the hazard from the toxic exhaust. Storage of cryogenics like liquid fluorine is easier. On the other hand, packaging a laser into a form that can be put aboard a satellite is difficult. Making it rugged enough to withstand launch and the rigors of operating in space is another obstacle.[31]

ASAT and DSAT possibilities

In many ways the ASAT laser is the easiest of laser applications. It calls for a relatively low level of power and a smaller main mirror since it operates in a vacuum. An orbital laser has a number of advantages over a more conventional ASAT system. It can strike at long range without the need for orbital maneuvering — just a straight line of sight from laser to target. Previous ASATs, either direct-ascent or orbital interceptors, can attack only one target at a time. The laser ASAT will wait until their respective orbits bring the targets into range. After destroying it, other targets could then be engaged until the laser was out of chemicals.

One design for an ASAT envisages a 200 000 watt laser combined with a 9.2-ft (2.8-m) main mirror (this is half the power of the NKC-135 ALL). The pointing accuracy would be 1 microradian or better which equals 63 in at 1000 miles (160 cm at 1609 km) range. Five battle stations are planned.[32]

Another study postulates a laser ASAT network which has to destroy 53 Soviet satellites. These are in three priority groups. Priority one satellites are to be destroyed within four hours of the outbreak of hostilities; priorities two and three within the first 48 hours. In addition the laser ASATs would be called upon to destroy four simultaneously launched replacement satellites and two more replacement satellites every four hours.[33,34]

A further development of this idea is the laser-defensive satellite (DSAT) which has a role similar to that of a fighter escort. Laser DSATs are an orbital patrol to protect against enemy ASATs, destroying them before they have a chance to attack. Another option is to equip the prospective target with a self-defense laser (instead of the infrared homing missiles envisioned in earlier scenarios). Weight of the installation is the main problem with this approach.

In the mid-1970s, United Technologies proposed a laser-defense system for B-52 bombers which would weigh 10 000 lb (4536 kg). Even assuming that the technology could have been transferred directly to a satellite, the weight would have been unacceptable: it constituted a third of the total payload of a Titan IIID. Essentially, such a system would comprise a DSAT laser and its fuel supply allied to an operational payload (eg, reconnaissance cameras, communications transponder, ELINT receiver, navigation system or early-warning telescope).[35]

A foreseeable use for an on-board laser is to blind the infrared seekers of an attacking ASAT. Only a comparatively low-power laser would be needed, perhaps of the same order as the 1 milliwatt laser found in school laboratories. Because of the large size of a fully fledged laser DSAT, it would most likely have to be a separate satellite watching over and protecting its unarmed charges.

A laser DSAT battle station is more demanding. For one thing, it would have to surpass the attacker's range. Evidently this calls for a 2 to 5 megawatts laser and a 13.1 to 26.3-ft (4 to 8-m) main mirror. Higher power and range in their turn call for an improved aiming system with an accuracy of 1 ft at 1000 miles (0.3 m at 1609 km), which probably means a laser operating on the radar principle. By this method, a low-power laser is fired at the target, the reflection is then detected, and finally the main laser fires. Estimates suggest that five to ten laser DSATs would be needed for an effective defense.[36]

There are at present no treaties banning either laser ASATs or DSAT weapons. The sole limitation upon their construction and activities is the SALT I ban on interference with reconnaissance satellites. Laser DSATs could act as 'enforcers' of free passage in space. The US has always contended that a satellite is not limited by the ban on over-flights of another country's air space. Space, like the oceans, is beyond territorial limits. No nation can arbitrarily determine what is or is not permissible.

A laser DSAT would have to cope with a number of different threats — from an orbital interceptor, a direct-ascent rocket or a 'space mine'. The last named might masquerade as a normal satellite, its true nature being revealed only when it suddenly maneuvered and attacked. A laser DSAT could deter any attempts at interference.

An operational laser DSAT demands an exact definition of 'attack'. At what point do a satellite's actions cease to be merely suspicious and constitute an attack? Rules of engagement must take into account the nature of the threat, the vulnerability of satellites and the realities of celestial mechanics. Maneuvering a satellite to make a close pass or illuminating it with a low-power laser such as that used in a laser radar are possible examples of illegitimate actions; especially if this involved several satellites simultaneously.

All nations engaged in space activities must know and understand 'the rules of the game' so that an innocent action is not misinterpreted as hostile. Rules of engagement like these serve to prevent friction when different military forces meet on the high seas or in the air during peacetime.[37] The classic example is provided by Soviet bombers and US aircraft carriers. Soviet bombers routinely fly out towards the carriers, Navy fighters meet them and pull into escort position, not overtly threatening but in position to take action. The crews wave, take pictures and, according to legend, hold up *Playboy* centerfolds. The bombers and their escorts fly past the carriers, being careful not to fly directly over them, then go home. The Soviets make the point that they can attack the carriers and the Navy reminds them that it can retaliate.[38] The destruction of the two Libyan aircraft in the recent past demonstrates what can happen when the rules are ignored.

Crucial questions for the laser to answer

Laser ASATs and DSATs *can* be built. The key question is: are they worth building? Does the added capability justify the probable huge expense? A laser ASAT must be more efficient than a conventional system in order to score. If the F-15 ASAT was just as good, a laser would be unnecessary. Further, does the protection that a laser DSAT can give other satellites justify the cost? Finally, how vulnerable are the lasers themselves to attack and what countermeasures are available to the target satellites? Might there be relatively cheap methods of disabling the lasers?

The anti-personnel laser has yet to find favor and for very good reasons. A laser machine gun or cannon could probably be built, but would be too expensive and complex for battlefield use. A laser is a line-of-sight weapon — in hilly terrain or woods, its range is short. Countermeasures are very easy; a soldier need only hide behind a rock or tree and lob a few mortar rounds. Even if they missed, dust would do the trick by coating the laser's mirrors. One simple countermeasure for the man in the field is a metal or mylar shield.[39]

Protective measures which the satellite can take are more complicated to achieve. As heat is the 'kill mechanism', countermeasures must protect the satellite's vital systems against its effects. A number of methods have been studied. The simplest is a highly reflective umbrella. Another approach is to coat the satellite with ablative materials similar to those used on spacecraft heat shields, which carry the heat away by melting. (The Space Shuttle, because of its heat-resistant tiles, would have a higher degree of protection than a satellite against laser attacks; it is essentially swathed in a heat shield.)

Again another possibility is to make the metal surface of a satellite so highly polished that it reflects laser light like a mirror, or to equip the satellite with corner reflectors. These might even reflect the beam back to its source with destructive consequences for the aggressor.

A multi-layer defense can be visualized: a coat of ablative material on polished aluminum backed by insulation or special design features to prevent heat transfer into the interior. Insulation cannot however be total or the heat from interior systems will be trapped and slowly fry the satellite.[40]

Electronics would have to be hardened against heating to reduce their vulnerability. Electronic filters can be added to infrared sensors to prevent burnout.[41] More effective radiators to dispose of the extra heat is one alternative. Careful choice of materials, particularly for optical systems, may also minimize damage.

Solar cells received early attention as laser countermeasures. Electrostatic-bonding techniques might be used to attach the clear covers to the solar cells, replacing the vulnerable glues presently in use. For attaching the cells to the array, new heat-resistant glues are required. Filter coatings could be selected which do not absorb energy at typical laser wavelengths. The wire connectors must be welded — solder melts at too low a temperature.[42] The vulnerability of solar cells to laser damage may encourage greater use of nuclear-powered generators in military space vehicles. (This method was in fact tested on the LES-8 and 9 communications satellites.)

Self-defense for satellites Until the 1980s, the danger of attacks by ground-based lasers was not taken very seriously. With the discovery of an apparently functional Soviet ASAT laser this changed. Steps were taken by the US Defense Department to provide reconnaissance satellites and early warning satellites, in geosynchronous orbit, with protection against laser attacks. ASAT counter-measures are now an important field. They have to provide protection against orbital interceptors and direct-ascent ASATs as well as space and ground-based lasers. This includes not only hardening against laser and nuclear effects but such means as electronic countermeasures, decoys, on-board defensive systems and a laser-warning device.[43]

Despite such efforts, there are limits to the defensive steps that can be taken. Countermeasures cut into a satellite's payload. Moreover, a sufficiently powerful beam will vaporize any material. With prolonged exposure, satellite systems will begin to fail, no matter how heat resistant they may be. To replace inevitable losses, the US has plans to orbit small, single-purpose satellites which could be launched from ICBM silos. They would operate for only a limited period of time during an emergency and communications is one use envisioned for them.

A similar proposal is called SOFAS (Survivable Optical Forward Acquisition System) and concerns infrared early warning probes. These would be launched if an ICBM attack was suspected and early warning satellites had been blinded. The probe is designed to scan the expected flight paths of Soviet ICBMs and its data will then be relayed to the President and Joint Chiefs of Staff.[44] Another safeguard against attack are the US' own 'Dark Satellites'. These are dormant satellites, placed in high orbits, difficult for ground tracking equipment to identify. They do not transmit and serve as on-orbit spares.

An example is the proposed Milstar communications satellite, meant to replace both the DSCS and SDS satellites. At least four would be placed into geosynchronous orbit and two or more in highly inclined elliptical orbits. An unspecified number of Dark Satellites may be parked in supersynchronous orbits up to 110 000 miles (176 990 km) high. If some of the geosynchronous satellites were destroyed, those held in reserve could be lowered to geosynchronous altitude to compensate. The Milstars' maneuvering capability would allow them to evade Soviet direct-ascent ASATs which require several hours to reach geosynchronous altitudes. The satellites would be spaced so that a single nuclear explosion could destroy only one. They would also have cross-orbit relay equipment enabling them to communicate with each other, reducing the need for vulnerable ground-relay and tracking stations.[45]

Countering the ICBM: past and present

The impregnable defense, so strong that an attacker can be battered into defeat at its gates, is a long-established military goal. In the age of nuclear weapons and ICBMs it is even more difficult to attain. The greatest difficulty in building such a defense against ICBMs is their high speed, since only 30 minutes separates an ICBM launch and detonation. Sea-launched missiles, depending on range and trajectory, can hit their targets in only ten or 15 minutes. To catch them requires a missile with an even higher velocity. The Sprint ABM had such a high acceleration within the dense lower atmosphere that an oxyacetylene torch was cool by comparison to its friction-heated skin.[46] Yet even this was not fast enough; valuable time was lost in accelerating the missile. As a laser beam travels at the speed of light this is not a problem. Even before the Sprint's engine was ignited, the laser would have hit its target.

The potential of a laser ABM was realized early. Soon after the laser's invention, the Air Force was funding research studies. These were part of Project Defender, an examination of novel ABM concepts. The power limitations that lasers had at the time prevented serious development.[47] Their potential was pointed out in a lecture by General Curtis LeMay, Air Force Chief of Staff and former SAC Commander, in a speech at Assumption College, Worcester, Massachusetts in March 1962. Talking about 'beam directed-energy weapons,' he said 'our national security in the future may depend on armaments far different from any we know today (. . . .) Perhaps they will be weapons that enable us to neutralize Earth-based ICBMs'.[48] Later, DoD censors removed scenes from an Air Force space film that showed a laser battle between US and Soviet spacecraft and the destruction of a Titan missile by a Soviet laser satellite.[49]

But could sufficient laser power be generated for this purpose? Development of the gas-dynamic laser was a beginning. Again, as with ASAT lasers, the simplest ABM laser would be ground-based with the laser replacing the interceptor missiles. A radar picks up the approaching warheads, their flight paths are calculated and the laser locks on and fires. As each warhead is destroyed, the laser switches to the next target.[50] This, however, suffers from the usual atmospheric problems compounded by the fact that the warheads are small and covered with an ablative heat shield.[51] Their size makes them difficult targets and even if hit by a beam, the heat shield will give them a measure of protection. The laser would have to lock on for a longer period to destroy them. Present Soviet re-entry vehicles can withstand 7000 watts per square centimeter per second (7 kilojoules/sq cm).[52]

There is also the problem of the terminal-defense environment. The warheads would be designed for minimum radar visibility. The multiple warheads from each ICBM would be mixed in with decoys, radar-jamming devices, chaff and assorted debris including deliberately fragmented boosters. The radars would have to sort through all this to pick out the real warheads. Inability to do so, with existing technology, has been the key reason for the failure of ABM systems to win approval.[53] The one advantage of a ground ABM laser is that it would not be limited in the number of targets it could attack, whereas a conventional ABM system can only attack as many targets as it has missiles.

A better approach would be to use the laser to destroy ICBMs during the boost phase, the critical period between launch and final-stage burnout. During this time a missile is particularly vulnerable and (what is more) there are no decoys to cope with. Most Soviet ICBMs and submarine-launched missiles are liquid fueled and have thin skins.[54] As the laser heats up the missile, it begins to lose structural strength. With protracted heating, the internal pressure and flight stresses (arising from acceleration, vibration, etc) would rupture the tanks, exploding the fuel and oxidizer.[55,56]

Solid-fuel ICBMs, such as the Soviet SS-16 and the US Minuteman, are tougher. Their major casings of fiber-filament reinforced plastic must withstand the internal pressure and high temperatures of the burning fuel. This could make them more resistant to laser attack. Other parts of the missile such as shrouds, interstage fairings, guidance compartment and wiring conduits are thin and vulnerable. The guidance and electronic systems, of both types, would be prone to heat damage.[57] A laser hardness of 0.5 to 1 kilojoules per square centimeter has been estimated based on the known skin thicknesses, materials and hoop stresses of Soviet ICBMs.[58] To defend against submarine-launched missiles, a laser-carrying aircraft is one possibility (both US and Soviet ICBM fields are in the interior and out of the range of aircraft). Soviet missile submarines patrol in three areas: between the US west coast and Hawaii, in the mid-Atlantic and off Norway's North Cape.[59] Laser-ABM aircraft could mount standing patrols over all these areas.

The laser battle station

A good deal of attention has been given to a much more exotic concept however — the laser battle station. One study ranked the technology necessary as being as important as the invention of the wheel, computers and nuclear weapons.[60] Although designs vary, such proposed battle stations would use large chemical lasers effective over thousands of miles with power measured in millions of watts. The beam would be controlled by a large main mirror. Guided by a precise aiming system, it may be held on a fast-climbing ICBM until the missile exploded. When this stage is reached, ICBMs could be swept from the sky literally in a flash, ending the reign of the ICBM and lifting its long shadow from the Earth. A network of these orbital battle stations might watch over the world, poised to prevent any attack and so guard the peace. Many different designs have been proposed for such battle stations. Certain features are standard. The battle station would carry some 40 per cent of its mass in the form of hydrogen and fluoride lasing gas in large insulated tanks.

A chemical laser can make very efficient use of the lasing gas. A single gram produces 500 joules; a 5 million watt laser firing for ten seconds would require 220 lb (100 kg) of fuel. A wide range of laser outputs has been studied. Typical are powers of 5 to 25 megawatts; between twice and over 10 times that of the Navy's Sea Lite laser.

Conventional laser designs are not adequate for these power levels, since a greater flow of lasing gas and larger optical cavities would be required. Previous chemical lasers have used a linear-nozzle design — shaped like a long, narrow rectangle. When the length of the nozzle is increased, the pressure of the fuel flow will distort it; this is avoided by using a cylindrical-nozzle design which increases the effective length without the associated problems although a cylindrical nozzle can create problems in permitting the laser to beam out.[61]

Nevertheless, since work began on cylindrical nozzles and combustion chambers in the late 1970s, they have become standard. Improvements in nozzle design have yielded increased power.[62]

Another standard part of laser design is a large main mirror. When the narrow beam emerges from the laser, it is reflected onto the main mirror. The mirror controls and directs the beam, expanding it to the necessary diameter in order to project it across several thousand miles. The slight curve of the mirror focuses the light, resulting in a condensed beam hitting the target.[63]

The mirror is not a single unit but segmented. Actuators adjust the mirror's segments, smoothing out any irregularities in the beam which are due to it heating the mirror. A computer senses the distortions, computes the necessary corrections and moves the actuators. Without these adjustments to the mirror, the beam would be ineffective.

The size of the mirror is determined by the wavelength of the laser: the shorter the wavelength, the smaller the mirror.[64] On the other hand, a shorter wavelength requires a more precisely figured mirror.

A battle station's mirror must be lightweight: such a mirror uses extremely low-expansion glass, the front side of each segment taking the form of a thin plate. The back is reinforced by glass ribs. After the parts are fitted together in manufacture, the mirror is heated, so fusing them into a single unit. The result is a strong but lightweight mirror. This technique is now commonly used to fabricate large mirrors for astronomical telescopes, including the mirror for the Space Telescope now in development.[65] The process can be both time-consuming and costly.

United Technologies Research Center has proposed making the mirror segments from a mixture of glass and graphite fiber reinforcements, coating the segments with vaporized silicone. This coating reduces the stresses on the graphite fibers due to heating. Analysis of the weight, optical characteristics and thermal properties of the composite mirror indicate it may be equal to or better than a conventional glass mirror. The composite mirror's materials are readily available, fabrication and polishing is easier and less expensive.[66] Several mirror sizes have been proposed — 7.9 ft (2.4 m) for a subscale demonstration model: 13.1 ft (4 m) for a simple ABM laser and much larger designs for use with super battle stations able to destroy ICBM warheads.

No matter how powerful the beam, it is useless unless accurately aimed. This task is made more difficult by the extreme narrowness of the beam and the great distances at which the engagement may take place. As already explained, to kill a missile the beam must hit, then dwell on the target until it is destroyed. The beam cannot wander about: such 'jitter' would spread the laser's energy over a wide area, lessening the damage and increasing the time the beam must hold the target.[67] Accuracies of up to 0.05 microradians are considered necessary. This is equivalent to 3.15 in at 1000 miles (8 cm at 1609 km).

In contrast, a nuclear ABM can destroy its target even if it misses by miles.[68] The aiming-fire control system of a laser would include sensors to detect the missile within 50 seconds after launch, precisely track it, select the aim point, fire the weapon and then keep the beam on target.[69] The system must be able to determine when the target has been disabled so that it can immediately switch to another target.

The aiming system on the Navy's San Juan Capistrano test laser uses a passive infrared sensor built by Hughes Aircraft. An operator points it at the target. When it is in the sensor's field of view, the system begins to 'rough track' the target. It then switches over to a more precise tracking system and the laser is automatically fired.[70]

For laser battle stations, a more advanced firing sequence is required. Initial aiming would probably be by infrared sensors tracking the missile's exhaust plume. The low-power beam of the laser radar then begins to scan the target area and when the beam strikes the missile, its reflection is detected. From the returns, the speed and direction of the target can be determined. Once the laser radar has locked on the main laser fires. Despite the fact that the laser beam travels at the speed of light, the long range and the target's own velocity of several thousand miles per hour means the battle station must anticipate its target path. The distance is however only two or three missile lengths. In contrast, an ABM missile must be fired at a point perhaps 50 miles (80 km) in front of an approaching warhead.

A 'closed-loop' aiming system has been tested by Lincoln Laboratories, codenamed Fire Pond, which employs a modest-power carbon dioxide laser and velocity-measuring equipment.[71]

Space lasers and existing ABMs

A study presented to Congress by four laser experts envisioned 18 battle stations in 1087-mile (1750-km) high polar orbits. To avoid any gaps in coverage there would be three orbital rings of six lasers each. Each battle station would have a 5 megawatt hydrogen fluoride chemical laser with a 13.1 ft (4 m) mirror. The lasers would be 19.7 to 26.2 ft (6 to 8 m) long, weigh about 37 400 lb (17 000 kg) and have enough fuel for 15 to 19 minutes firing. Two or three Shuttle flights would be required to establish each station with orbital assembly. The core unit, with the laser and battle-management equipment, would go up on the first flight; subsequent flights would carry the fuel tanks.[72]

Another study indicated that only one quarter of the laser battle stations would be in position to neutralize the missiles, the remainder being needed to defend against submarine-launched missiles.[73] The battle stations would have only 300 seconds to destroy each ICBM. This is the time between the missile leaving the atmosphere and the burnout of the final stage. At the start of the engagement, the lasers are working at long range. As the attack develops, some of the ICBMs are destroyed closer to their targets. (Typically, the beam destroys an ICBM in less than ten seconds. This varies if the combination of the missile's trajectory and the laser's orbital path cause a major change in geometry; the illumination time might increase to 20 seconds.) Such a network is envisioned as destroying ICBMs at the rate of one per second. The study considered a number of attack scenarios, from a simultaneous launching of the entire SS-18 force to firings spanning 15 minutes.[74]

A 5 megawatt-4 meter diameter laser is the typical design for these studies and numbers range from 18 to 30 battle stations, which could engage as many as 3000 targets.[75] The goal is damage limiting; reducing the number of missiles that get through rather than stopping them all. Under the damage-limiting plan, the laser battle stations would be the first layer of a three-phase ballistic missile defense backed up by non-nuclear homing missiles for mid-course interception. The final phase would be a low-altitude terminal defense system with the object of destroying the warheads after they had re-entered the Earth's atmosphere.

This three-level approach is intended to ensure maximum protection. The missiles which escape the laser network next face the mid-course interceptors. The surviving warheads are met by the terminal defense.[76] The lasers alone would be expected to stop several hundred ICBMs.

A completely laser-based damage-denial system is another matter. This would be designed to stop *all* incoming missiles; a shield against the ravages of nuclear war. This concept is the one commonly thought of in connection with laser battle stations. By it, lasers would largely replace conventional ABM systems. Estimates of the number of lasers needed for damage denial range from 20 to 80 lasers.[77] One estimate investigated a 100-satellite network of 25 megawatt -15 meter diameter battle stations. Another study describes a 98 ft (30 m) main mirror matched by a 60 megawatt laser![78]

So much for the dream, but the reality is more complicated. A problem which arises at once is the launching of such lasers. A laser with a 13.1 ft (4 m) diameter can fit into the Shuttle's payload bay. The larger designs cannot. Either the battle station must be assembled in space from components brought up by multiple Shuttle launches, or the lasers must be launched as a single unit aboard a heavy-lift vehicle. This could be developed from Shuttle hardware (solid rocket boosters, external tank and main engines) to launch large, outsized cargo. Such boosters have been proposed and are being actively studied.

A Soviet heavy-lift vehicle — the so-called G booster — is under development. Its payload is estimated to be 297 624 lb (135 000 kg) for low Earth orbit.[79] Its roots can be traced back to the early 1960s and the Soviet manned Moon-landing program. The first exploded on the launch pad in 1969, and two others suffered launch failures in 1971 and 1972. After these mishaps, the program was reduced to a research and development effort. This low level of activity continued until the late 1970s when serious development work resumed. Latest estimates suggest that a new type-G could be operational by the mid or late 1980s, always assuming it avoids the misfortune of its predecessor.[80]

Space cruisers and orbital lasers
The battle stations themselves represent a substantial investment; perhaps a billion dollars each.[81] To justify the expense, they would need to have a long operating lifetime. Periodic maintenance, system updating and reprogramming would be required.

This in its turn introduces another major developmental expense: a small, manned utility spacecraft to service the battle stations. An SRI International study envisaged that this space cruiser (as it is called) might take the form of a winged spacecraft about 22 ft (6.7 m) long and weighing some 5000 lb (2268 kg). A variety of launch vehicles were considered, including the MX ICBM and Space Shuttle. Three space cruisers could be carried by a single Space Shuttle. There was even the possibility of air-launching a single vehicle from a modified Boeing 747.

For orbital maneuvering, use could be made of an RL-10 Centaur engine with strap-on hydrogen and oxygen tanks. At present however, because of Shuttle weight limitations, the space cruisers would not carry a full load of propellents in their strap-on tanks. Orbital refueling of one kind or another would be necessary. For low-orbit operations, this is not a problem. With full tanks, the space cruisers would have a velocity changing ability of 28 000 ft per second (8534 mps). With this maneuverability, satellites from low Earth orbit to geosynchronous altitude are within serviceable range. An estimated $1 billion stands to be saved over a period of several years by using such a system.

Besides orbital maintenance, the space cruisers could be used for reconnaissance, ELINT or communications. (They would not be armed, however). After the space cruiser had completed its mission, the strap-on tanks would be separated, the spacecraft retro-firing and landing on a runway. After servicing, it flies again.[82]

Other roles besides ASAT, DSAT and ABM missions have been proposed for laser battle stations. Aircraft spend much of their time at high altitudes because jet engines are more efficient at low atmospheric pressures. Bombers fly at high altitude to increase range. With 10 to 60 megawatt lasers overhead, they would be vulnerable.[83]

Space lasers have been suggested for air defense of the continental US. Their task would be to destroy 176 Soviet bombers in the six hours before the latter came within 1200 nautical miles (2222 km) of the US coast. These lasers may also be used to suppress Soviet air defenses before the arrival

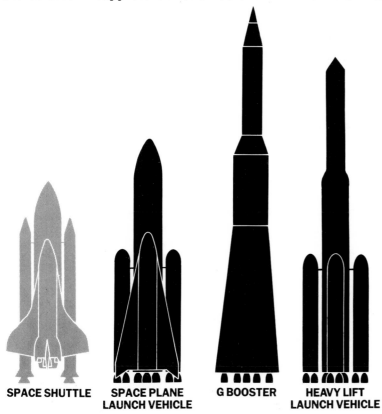

SPACE SHUTTLE   **SPACE PLANE LAUNCH VEHICLE**   **G BOOSTER**   **HEAVY LIFT LAUNCH VEHICLE**

Soviet superboosters. A new family of space rockets is now being developed to carry Soviet space ambitions into the 21st century. All use liquid propellants and are larger than the US Space Shuttle. The Soviet Space Shuttle launcher has a liftoff thrust of approximately 6 000 000 lb (2 721 600 kg); a liftoff weight of 3 307 500 lb (1 500 000 kg); and a payload to 111 miles (180 km) orbit of 132 276 lb (60 000 kg). The more powerful heavy-lift launcher has two to three strap-on boosters and has a liftoff thrust of 9 000 000 lb (4 082 400 kg). Its payload to 111 miles orbit is about 330 750 lb (150 000 kg). The figures are those given by the US DoD (Soviet Military Power 2nd Edition); they appear optimistic in terms of anticipated Soviet Space Shuttle capability. The G-type booster which failed at Baikonur in 1969–1972 is also represented.
Evidence from satellites shows that new constructions at the Baikonur cosmodrome include a long runway north of the launch pads for the superboosters. It is probable, therefore, that the US will face a serious Soviet challenge by the end of the 1980s.

A photograph taken over the Baikonur cosmodrome by Landsat 3 at 131 ft (39.9 m) resolution shows new constructions for the forthcoming Soviet superbooster. A major landing strip is at the northern end, which may be used to fly in upper stages of large space rockets on the backs of modified Bison bombers. Later the strip may be used for recovery of the Soviet 'Kosmolyot'. The location of the original Sputnik/Vostok pad (photographed by a U-2) is also indicated.

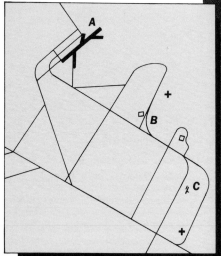

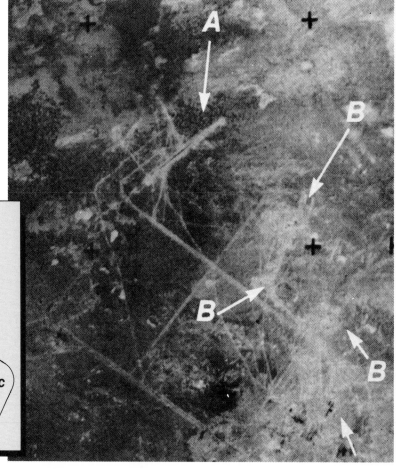

A Runway.

B Apparently the superbooster launch pad under construction.

C Large assembly building for new superbooster and new construction area, larger than the basic Tyuratam launch pad area.

of US bombers, their targets including Soviet interceptors and airborne-warning and control aircraft.[84] At lower altitudes, aircraft are less vulnerable as a result of the protective blanket of the atmosphere. This may also rule out a laser-bombardment system attack on surface targets.

Laser battle stations: the technical problems

Only the Space Telescope at the moment has the accuracy which a space-based laser requires. The telescope is designed to have an accuracy of 0.025 microradians, which is equivalent to 1.5 in at 1000 miles (3.8 cm at 1609 km). The difference is that the Space Telescope has no need to stabilize swiftly on its target, since acquiring a star or galaxy can be a leisurely process. A battle station in sharp contrast must lock on a target in a few seconds. Additionally, the Space Telescope is not very active during operations. A filter wheel may turn or the solar panels may change orientation but none of this activity compares with the firing of a chemical laser. Sighting accuracy is problematic.

Critics have also said that the exact location of several thousand fixed ICBM silos would have to be programmed into the battle station's computers. This list would have to be regularly updated. Further, as each laser passed out of the battle zone, its targets would be handed on to the next station in line, the implication being that a mobile ICBM would defeat the system.[85] An examination of published sources does not in fact support this. The battle stations would be self-targeting. On-board systems look for the missiles, acquire them, point the weapons and track the target, destroying even submarine-launched missiles in this way.[86]

One of the more obvious vulnerabilities of orbital nuclear weapons is their command and control network. For laser battle stations, this is a major political and military problem. The lasers have only four or five minutes to destroy a particular missile, which allows little time for political consultations or the transmission of the firing authorization. To be effective, the battle stations must begin firing as soon as an attack is detected. Two political options present themselves, and both are uncomfortable.

First, that the lasers are pre-programmed to automatically open fire on detection of attack. The political leadership of the country would then turn the power of decision over to a machine; not a very popular idea.[87] What is more, any such design is sure to come under fire from the 'berserk computer' school who would undoubtedly invoke the possibility of a malfunctioning battle station indiscriminately blasting away at any satellite, missile, airliner or flock of geese its targeting system detected. Even if the system was specifically designed to eliminate this possibility, the image lingers of a mechanism which might set off an international crisis if it malfunctioned.

The second option is that the battle stations be controlled from the ground or by a manned space station. The authority to fire would in this case be assigned to an individual on-site — the ADC Commander, the Station Commander or a Lieutenant sitting in front of a control panel. Political difficulties would of course remain in allotting so much power to an intermediary, and communications would be vulnerable to countermeasures into the bargain.

In view of the difficulties, are laser battle stations worth building? As we have seen, space bombs were eliminated from consideration (at least in part) because of their vulnerability to attack, and considerable attention has been given to ways that a network of laser battle stations could be neutralized. The most spectacular method is the electromagnetic pulse attack, in which a large nuclear weapon is exploded at high altitude. The electromagnetic energy generated destroys the battle station's electronics with a massive power surge and indiscriminately puts satellites out of action — perhaps even those in geosynchronous orbit, though this area is not well understood and harbors many unknowns.[88] The attacker could then launch his missiles unmolested. It is important to remember that EMP damage to friendly satellites was suggested as the 'real' reason for Program 437's abandonment even though *available* documentation does not apparently confirm this.[89]

Another problem with an EMP attack is uncertainty over the pulse's maximum power. The only hard evidence is from the Starfish nuclear test on July 9, 1962. In Honolulu, some street lights went out, power lines shut down, circuit breakers popped out and burglar alarms were tripped. The effects, however, were highly sporadic. Telephones still worked and there was no mass destruction of transistors or other sensitive solid-state electronics.[90]

Also, it is possible to harden electronics against EMP. These systems are designed to detect the surge and protectively shut down the electronics until it is past. They act faster than the pulse can 'kill'. The satellite would have to be re-programmed from a storage protected against harmful radiation. Only this programming would survive the EMP; any recent programming without this protection would be lost during shutdown. It is not possible to protect everything — only the most important systems and programs. Design factors in this connection include avoiding long straight-cable runs and using optical fibers and vacuum tubes, which are a million times less vulnerable than transistors. Such hardening is meant to avoid the so-called 'cheap shot' that could wipe out multiple satellites.[91] Existing test systems cannot exactly reproduce the conditions of a nuclear explosion and this, alongside the secrecy which cloaks the efforts to meet the EMP threat, means that it is difficult to be more precise about the system's vulnerability.

Attacking a laser battle station

EMP aside, there are more conventional ASAT threats. The laser battle stations would have to be protected against the flash and radiation of a nearby nuclear explosion.[92] Attacks by other lasers, both ground and space-borne, must be taken into account as well.

Another method of destroying a laser battle station is a mass attack by 20 or so F-15-launched ASATs. The miniature vehicles would be hard to detect and defend against. Even if some were stopped, enough would get through to destroy the target.[93] One obvious countermeasure is altitude since the higher the target, the larger the booster needed. Launching could thus be detected and dealt with in the boost phrase. With the battle station's long range, an increase in orbital altitude would probably not significantly affect its ability to destroy ICBMs. Whatever method an attacker used, the goal remains to tie down the lasers, keeping them busy with repeated attacks. In this way, some lasers would be destroyed, others damaged.

This would open holes in the defensive line through which the ICBMs could be launched. Since attacking the defense is such an obvious tactic, war games have been played to find out how vulnerable a laser battle station would be. The postulated weapons included Soviet orbital interceptors, new Soviet ASATs, space mines and Soviet lasers. The conclusion was that, despite casualties, enough battle stations survived to make the concept feasible.

The stations are to a certain extent self-protecting. The studies further concluded that cheap countermeasures were not very effective, and that even extensive and very costly attack profiles did not guarantee neutralization of the laser battle stations.[94]

Could ICBMs in the meantime be protected against laser beams by coating them with ablative material? In theory, this would keep the missile safe until the boost phase was over, but would involve a weight penalty, reducing the range and/or payload. Subsequent generations of ICBMs may take these factors into consideration from the start, survivability becoming an important consideration.[95]

Another countermeasure, which does not involve such weight penalties, lies in 'spinning' the booster so that the attacking laser smears its energy over the surface and does not dwell on one spot. To destroy the missile, the laser must pump more heat into the ICBM than the missile sheds while spinning and consequently must hold the target for a longer period. The number of ICBMs each laser can destroy is reduced as a result. The effects of rapid spinning on booster-guidance systems has not yet been documented.

Another procedure is to polish the missile, turning its surface into a mirror. A General Accounting Office study assumed 24 battle stations at an altitude of 745 miles (1200 km) in three orbital rings. The pointing accuracy is 0.2 micro-radians. In a 15-min salvo, calculations suggest this network could destroy about 500 missiles carrying 3000 warheads if they had a normal finish. However, if the missiles' skins were polished this drops to 175 missiles and only 1000 warheads.

This finish would need regular maintenance and could suffer from reflected exhaust gases, aerodynamic heating and debris thrown up during launch.[96] It would also have to reflect the wavelength of the laser, for what appears shiny in visible light may not do so in infrared. This procedure would only work for missiles with bare-metal finishes (like the now-retired Titan II). Minuteman missiles could not at present be treated in this manner since they are covered with heavy-guage aluminum, cork and titanium to protect them against dirt and debris from nearby nuclear explosions.[97,98] Other missiles are covered with camouflage paint. The hardening of missiles means higher-power lasers, larger mirrors, and more extensive orbital networks, which increases development and deployment costs enormously.

Work in progress

The US has been working on three projects related to space lasers. Together they cover all of the elements necessary for an orbital battle station. The first of the triad is Alpha, a project to develop a 2 to 3 megawatt hydrogen fluoride continuous-wave laser. Its purpose is to determine if a laser can be scaled up to higher-power levels. Alpha, which uses cylindrical combustion chambers, is designed to be expanded by the addition of generator modules. It is estimated that it could achieve up to 10 megawatts with reduced fuel efficiency. Ground tests are planned in the mid-1980s.

The problems of the main mirror will be explored in the Large Optics Demonstration Experiment (LODE). This involves a 13.1 ft (4 m) main mirror and beam-control system and includes the adaptive optics needed to maintain beam quality. The ground-based, low-power demonstrations are also planned for the mid-1980s.

The pointing and tracking questions will be addressed by Talon Gold. This is a space-based demonstration of a scaled-down acquisition, tracking and aiming system combined with a low-power laser. Use will be made of high-altitude aircraft and spacecraft as tracking targets and for data on thermal signature and background. Design objectives of the Talon Gold laser are to track a target at ranges up to 932 miles (1500 km) with an accuracy of 0.2 microradians. Launch will be during the mid-1980s aboard the Space Shuttle.[99,100]

All three programs are still in the early stages. Public information about them is very limited, amounting to virtually only their names and purposes. The 1980s will be a decisive period for laser weapons. By the end of the decade, the US and the USSR will have hard data on these questions. Has the time come for such weapons?

Although the triad projects are important, many of the answers will come in a piecemeal fashion: a new nozzle design, a refined chemical ratio, modified optical chambers and mirror configurations and higher-capacity electronics — all important but none of them newspaper-headline developments.

It had been proposed that the NKC-135 ALL be used to shoot down a Polaris missile launched from a submarine off Kwajalein Atoll. The aircraft would fly at 35 000 ft (10 668 m), some 40 miles (64 km) from the submarine. Next the NKC-135 will shift to Vandenberg AFB where it would attempt to destroy a Minuteman missile.[101] The importance of such a test cannot be overstated. However, three basic problems have seriously affected the test program: air flowing around the laser turret, tracking difficulties and the effects of aircraft vibration.

Chapter 6

PARTICLE BEAM WEAPONS

'We should abandon this immoral and militarily bankrupt theory (...) and move from Mutually Assured Destruction to Assured Survival (...) Should the Soviet Union wish to join in this endeavor (...) we would, of course, not object (...).'

General Daniel O Graham

Research into energy beam weapons is not confined to lasers. Particle beam weapons are also being developed. Lasers and particle beam weapons are alike in that both use a narrow beam of concentrated energy to destroy a target. Beyond this superficial similarity, however, the two are very different. Particle beams have completely different operating principles and accordingly have different advantages, limitations and demands. A laser relies on the exotic effects of quantum mechanics which lie outside everyday experience. This is not true of particle beams whose basic principles can be demonstrated with ordinary magnets.

Anybody who has played with magnets has noted that like ends of magnets repel each other. If the magnets are large enough, a considerable amount of force is needed to press them together. When the pressure is released, they fly apart again. With enough ingenuity, one magnet can be made to 'float' above another, since electromagnetic repulsion provides enough force to overcome the pull of gravity.

The principle demonstrated by the magnets — namely that like charges repel — is employed in a particle beam weapon to accelerate electrons or protons to high energies and velocities. Electrons are negatively charged. Protons — which make up the atom's nucleus — carry a positive charge.* The hydrogen atom is a single electron orbiting a single proton; the electrical charges balance out, so that there is no net charge. The neutron, which together with the proton makes up the nucleus of more complex atoms, is electrically neutral as its

*Protons (a part of the atom) should not be confused with photons (the basic unit of light).

name suggests. This means that it cannot be used in particle beam weapons.

Curious as it seems, particle beams have a long history. They were first used in early research into nuclear physics in which high-energy beams of subatomic particles were used to probe the structure of the atom. The first linear accelerator was developed by Ernest Lawrence in 1931. Since then, particle accelerators have burgeoned, becoming complex and sophisticated technological centers. The fact is that most homes contain an application of Lawrence's device in the TV set.

The first part of an accelerator is a particle source. Electrons are easy to produce: a high current is simply sent through a wire filament to boil off the electrons. This is exactly the procedure used in a TV picture tube. For a beam weapon, the major difference is that a much higher particle density is needed. One procedure used to achieve this is to ionize hydrogen gas with an electrical current. In a process which can be adapted for either proton or electron beams, electrons are stripped off in this fashion, and separated from their protons.

A more sophisticated device is a negative hydrogen beam, by which an extra electron is added to a normal hydrogen atom. As a result of a technique which entails sending an electron beam through hydrogen gas, two electrons and one proton are produced, with a net negative electric charge. This artificially produced negative hydrogen atom is stable and because it is negatively charged it can be magnetically accelerated.

The White Horse project

Such are the bare bones of the procedure used in the White Horse project, a research effort aimed at developing a particle beam for use in space. White Horse was formerly known as Sipapu, an American Indian religious word meaning 'Sacred Fire', but this title was dropped for fear of offending the American Indian lobby. Work is in progress at Los Alamos, New Mexico where the first atom bomb was developed. When the particles are emitted from the source, they are moving at very low velocities. This can be a problem as magnetic focusing depends on velocity; the slower the particles, the stronger the magnets must be. Unfortunately, the first stage of the accelerator has no room for powerful magnets since the velocity of the particles must be raised as a first step and this requires large, complex generator systems and power sources. The particles must also form a bell-shaped bunch and be properly spaced in order to enter the accelerator. (In the past, these bunches have been poorly formed, with long tails and a considerable amount of work has been devoted to this area.) The solution is a device called the Radio Frequency Quad-

Particle Beams and Linear Accelerators

Scientists are still debating the merits of charged particle beams and whether or not they will be effective as weapons against aircraft, missiles and satellites. In principle what is needed is a special kind of atomic accelerator in which streams of subatomic 'bullets' (protons or electrons) are accelerated by means of precisely tuned electromagnets, which drive the particles to speeds approaching that of light.

Below: A synchrotron is fed with speeded up particles from a linear accelerator. Accelerating units in the synchroton ring increase particle speed with each circuit. When the required velocity is reached, the particles are diverted out of the device to the target.

Above: The Lawrence Livermore accelerator in action. This laboratory is also reported to be investigating the controversial X-ray laser.

Right: As a charged particle enters a drift tube, the electrical charges are switched in such a way as to keep the particle moving forward. The process is one of attraction and repulsion. Each time a particle crosses the gap between drift tubes, it is accelerated.

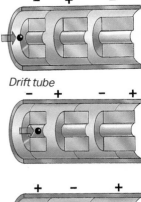

Drift tube

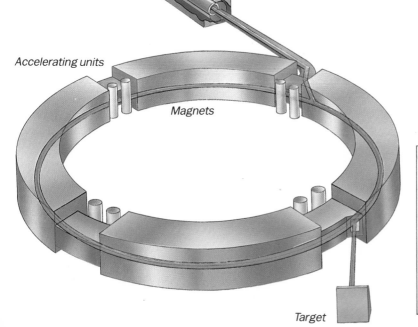

Linear accelerator

Accelerating units

Magnets

Target

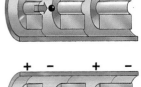

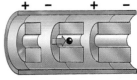

Circulating in Washington in 1983 was a Congressional report which stated that Soviet scientists have been tasked to concentrate on the 'compact formation of plasmas in a state capable of electromagnetic acceleration'. The result could be an immensely powerful plasma gun accelerating toroidal plasmas of free subatomic particles – a potentially fearsome weapon.

Future Beam Technology

This gallery of future beam weapons may become practicable by the turn of the century, ready to blunt any ICBM attack.

A chemical laser battle station disposes of a Soviet ASAT. Operating at about 5 megawatts, it requires a 13 ft (4 m) mirror. Short-wave systems require smaller mirror optics.

The X-ray laser — with 50 laser rods trained on as many ascending ICBMs — forms the first wave of the ABM defense. The laser is itself consumed by the nuclear explosion it generates, but by then the beams are streaking to their targets.

Space mirrors refocus laser energy from high-elevation ground stations against surviving ICBMs. These orbital mirrors, launched when attack seems imminent, form the second wave of the ABM defense.

Weapons based on particle beams (electrons, protons and ionized particles) also join the battle at the last moment in an attempt to neutralize ICBMs which have slipped through the earlier screens.

Peaceful applications of beam technology include huge solar power stations in geostationary orbit which convert sunlight in space to electricity. The energy is beamed to Earth receiving stations as microwaves, where it is transformed into direct current for use in homes and factories. At times of peak demand power relay satellites distribute electrical energy from one country to another. Microwave energy (and hence electrical power) may be transferred between space vehicles, even directed to scientific bases on the Moon. New space-propulsion systems may depend on receiving controlled amounts of laser energy beamed at them from space stations, like the Space Shuttle here.

Laser battle station

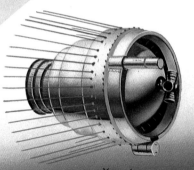

Orbital mirrors

Soviet anti-satellite

X-ray laser

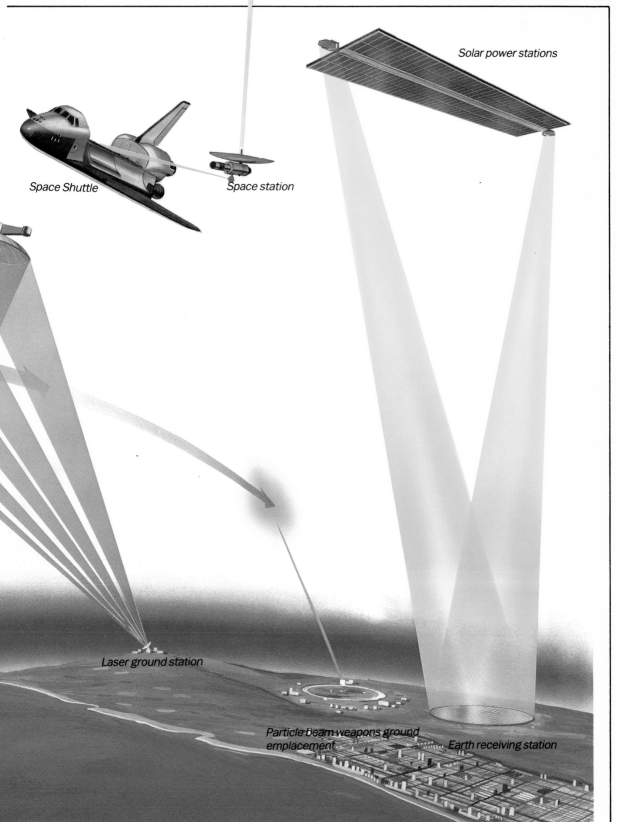

Solar power stations

Space Shuttle

Space station

Laser ground station

Particle beam weapons ground emplacement

Earth receiving station

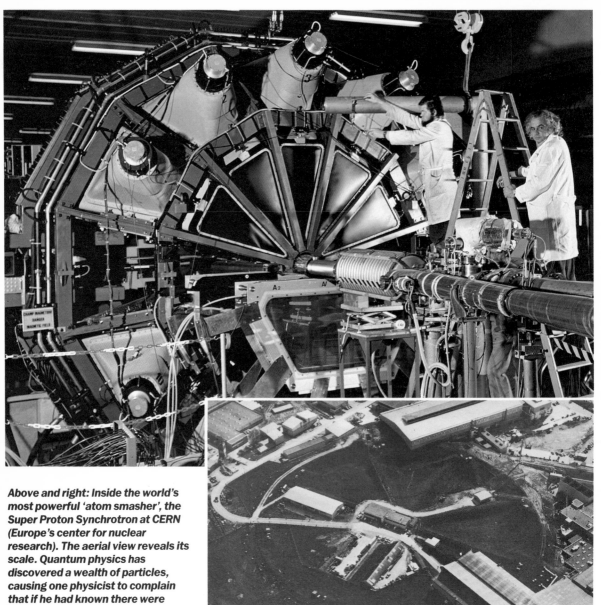

Above and right: Inside the world's most powerful 'atom smasher', the Super Proton Synchrotron at CERN (Europe's center for nuclear research). The aerial view reveals its scale. Quantum physics has discovered a wealth of particles, causing one physicist to complain that if he had known there were going to be so many he would have studied zoology.

rupole injector (RFQ) which focuses, bunches and accelerates the particles using only magnetic fields. The RFQ is not limited by particle velocity and so removes the need for the earlier complex power supplies.

Interestingly, US work on the RFQ was begun in 1978 by the White Horse team after sketchy descriptions of such systems had appeared in Soviet scientific papers. Over the next two years, the concept was developed, computer programs were written and finally the test version was built.

Initial tests made in February 1980 proved successful. The Radio Frequency Quadrupole is only the first step of the acceleration process, serving to simplify the design and to prepare the particles for entry into the main accelerator.[1]

White Horse is a linear accelerator using a series of devices called drift tubes. Drift tubes are short metal tubes running in a line down the center of a vacuum chamber with a gap between each one. The beam travels in a straight line through them. These drift tubes are connected to an oscillator which changes their electrical charge at a fixed rate.

The best way to illustrate the method is to follow the charged particles down the accelerator. The drift tube lying ahead of the negative hydrogen bunch has a positive charge and so it attracts the particles. The drift tube behind them is negatively charged, so it repels them.

As the particle enters the drift tube, its constituent particles are electrically isolated. As there is no net force on them, they coast. While they are passing through the drift tube, the electrical charges switch from positive to negative. By the time the particles leave, the charge has stabilized as negative and the drift tube thrusts them onward by a process of repulsion. The next drift tube in line pulls them forward again by a process of attraction and so the cycle continues. Each time the particle bunches go through the gap between the drift tubes, they are accelerated to a higher level. Because the oscillation cycle is at a fixed rate, the drift tubes are progressively longer to compensate for the increased velocity.[2] As speed increases, the effects predicted by Einstein in his General Theory of Relativity start to occur. The particles gain mass. As they do so, more and more energy is required to speed them up. This steadily becomes a matter of diminishing returns. It is one of the consequences of Einstein's theory that even given infinite resources of energy, the particles could not be accelerated to the speed of light, their velocity always falling short by a fraction.

For the accelerator to function properly, the timing, spacing and bunching of the particles must be correct. The drift tubes must switch charges as the particles pass through them, and if this occurs too early or too late the particles will decelerate. Furthermore, if the spacing of the bunches is not carefully co-ordinated with the switching time, the beam will be adversely affected. A third complication is that the shape and density of the bunched particles is also important. The development of the RFQ, which facilitates all these things, was consequently a major breakthrough.

The design of the linear accelerator will vary according to what particles are in the beam. An electron accelerator requires less energy than others because its particles are smaller. Protons and negative hydrogen require more energy because their particles are more massive. Nor is the linear design the only type of particle accelerator in existence.

Researchers use a number of designs under the general name of cyclotrons because the particles are sent in circular paths. Although details vary, all these designs use magnetic forces to bend the particles into the required path and during each trip around the circle, the particles receive an extra boost. Another striking feature of these devices is their size. The CERN proton synchroton near Geneva, Switzerland is 0.9 miles (1.5 km) in diameter.[3] The national accelerator at Batavia, Illinois is 4 miles (6.4 km) across.

A particle beam's power is expressed in electron volts and amperes of current. An electron volt is defined as the energy necessary to accelerate one electron through a potential of 1 volt. (This indicates the energy of the beam in much the same way as water pressure indicates the force of water flow in a pipe.) For beam weapons, energies of one billion to ten billion electron volts are necessary.[4] Current in amps is also important and for a weapon this must be in the tens of thousands. This measurement gives the 'flow rate' of the particles and indicates whether the particles are coming out of the accelerator slowly or all at once, much like emptying a large water tank with a garden hose (low amperage) as opposed to splitting its sides open (high amperage). Interestingly, the power of a beam with a low electron voltage and a high current may match that of another with a high voltage and low current.

Beam weapons and TV ray guns
A charged particle beam is aimed magnetically. This is done by electrical currents, allowing the beam to sweep from one location to another almost instantaneously. A simple example is the TV cathode ray tube. Its low-power electron beam sweeps across the 525 lines that make up the picture 30 complete times a second. Nothing mechanical could move back and forth at such a rate. Once the charged particles leave the accelerator, however, a problem appears. The beam is now a mass of particles with a like charge. The particles mutually repel each other and as a result the beam begins to fly apart, spreading and weakening. According to one estimate, a beam might expand to 1500 times its original diameter in the first 622 miles (1000 km) it covered.[5]

It may be difficult to provide the large amount of electrical power necessary for a space-based particle beam. The beam would need a higher current, so that although it had less energy it would contain more particles.

Another problem a space-based beam weapon faces is the Earth's magnetic field. This is expected to cause considerable bending of the beam, perhaps amounting to several degrees.[6] White Horse uses a beam neutralizer to avoid these problems. As the negative hydrogen atoms leave the accelerator, they pass through an exchange cell which strips off the excess electrons, removing the negative electrical charge. An electrically neutral beam of hydrogen atoms is left — hurtling towards the target at nearly the speed of light. Since the beam is now electrically neutral, the particles do not repel each other and are unaffected by the Earth's magnetic field.[7]

Once inside the Earth's atmosphere, things are straightforward. A hot electrically conducting channel is created as an electron beam moves through the air. It is predicted that the moving beam's own internal magnetic field will hold it together. If this proves true in open air tests, an electron beam for ground-based use is feasible.[8]

A proton beam encounters more serious problems in the atmosphere. As it penetrates the air, the protons will strip off the electrons from the surrounding atmospheric gases to form hydrogen, which consists of one electron and one proton. The hot hydrogen gas then combines with oxygen to form water, effectively transforming the 'death ray' into a water pistol. By the time the beam reaches space, it would be 100 000 times weaker than when it left the accelerator.

The answer may be to heat a thin column of air to several thousand degrees by means of either a laser or the accelerator itself. The proton beam could then be fired and enjoy a smooth passage through this channel of ionized air.[9] Atmospheric uses for particle beams are being investigated in the Navy's Chair Heritage program at Livermore Laboratory in California which is sponsored by the Naval Surface Weapons Center. If the weapon can be scaled down, it might be used to protect carriers against aircraft and missile attack.

The weapon's destructive effects
A particle beam destroys a target from the inside, quite unlike a laser. When particles enter the material, they cause several types of reactions. Their magnetic fields tend to pull atoms apart and bear them along, tearing holes in the target's atomic structure. Metals derive their strength from their regularly spaced atoms, an arrangement in which atomic electrical fields reinforce each other. As these bonds are broken, the structure is disrupted and the target's material weakened. This ripping effect can generate internal heating at the same time. The interaction of the magnetic field causes the particles in the beam to lose energy.

This lost energy is transferred to the target's atoms which are consequently dragged out of their normal position and made to vibrate. The tremendous energy needed to speed up the particles in the beam is in this way converted into heat over a large area. Such heating is all the more devastating since it occurs within the metal rather than outside, as in the case of a laser attack. The 'kill' is achieved by the beam's powerful impact. A shattering shock wave runs through the material, particularly if the particle density is high.

Yet another destructive side effect is the actual splitting of atoms. A proton may hit the nucleus and split it. When this happens to metals with low atomic numbers, such as aluminium, there is no atomic chain reaction along the lines of the uranium used in nuclear bombs. Instead, the two halves of the split atom are converted into other elements, such as nitrogen. This means of course that they are no longer structural materials and the target is weakened. Spectacular as this sounds, it is a rare event, as a nucleus is very small and the space between atoms very great.

When particles are slowed by interaction, they can lose energy in the form of X-rays and gamma rays. Known as 'braking radiation', this effect is especially pronounced with electrons because of their higher energy.[10] In addition to damaging materials, such radiation may erase programming in computer memories and even cause premature fission in nuclear warheads.

Even low-power particle beams can damage the semiconductors in computers. Semiconductors contain a carefully controlled amount of impurities within the semiconductor material, added by a device called an ion implanter: a low-power electron-beam system half the size of a living room. A beam-weapon attack, in addition to heating up the components, means an uncontrolled influx of contaminates. On this principle, which will radically alter the components' electrical properties, the beam could 'kill' the electronics of a missile, rendering the guidance system 'mindless'. In many respects, a particle beam weapon resembles a nuclear bomb with more narrowly focused and localized radiation effects.

There are various countermeasures available to defend against laser beams: polishing, ablative coatings, corner reflectors or spinning as explained. None of these work against particle beams and there is at present no known way of lessening their destructive effects.[11]

The speed of the attack is another advantage. A laser must be focused on a target for anything between one and 20 seconds. A particle beam 'kills' in fractions of a second; moreover, because it does not have to hold on a target as long as a laser, it could destroy more missiles in the same period of time.[12] This suggests that a defensive network of particle beam weapons would be better able to cope with a full-scale attack from incoming missiles than a laser system.

Designing an orbital beam weapon Although lasers and particle beams use different physical principles, they share certain limitations. Each must be aimed with precision and can only destroy a target by directly hitting it. The only real difference is the time each has to linger upon the target. Like laser battle stations, particle-beam weapons have to meet the threat of attack by F-15 ASATs, orbital interceptors, direct-ascent ASATs and other battle stations. The most serious threat, of course, might be from enemy particle beams.

A particle beam has problems all of its own. Beam instability or 'hosing' is one of them. This may arise from a number of causes: for example, interaction between the surrounding electrical fields and particles, or collisions between molecules and particles (depending on the type of particle source in use). Such instability causes the beam to depart from its normal path, wandering and whipping. Another problem is the beam's pulse rate. Five to 50 pulses per second are required if a beam weapon is to be effective.

Scientific research involving particle accelerators strives to liberate the various subatomic particles of quantum physics: quarks, mesons, hyperons, leptons, antimatter, and others. When, for instance, a beam hits a tungsten target, the resulting subatomic particles leave tracks as they pass through a bubble chamber filled with vaporizing liquid hydrogen. These tracks are then photographed and the particle behavior analyzed in detail.[13] In weapons research, on the other hand, interest has centered on the physics of the beam, its instability, how the beam affects matter and such technological developments as the Radio Frequency Quadrupole. Rate of fire is also very important. The CERN synchrotron has a firing rate of one burst every 8 seconds. An experimental weapons-test accelerator by way of contrast can produce 5 pulses per second, and a burst rate of about 20 pulses is evidently achievable.[14,15]

Consider as well that a particle beam weapon needs large amounts of electrical power to operate. For a ground-based weapon this is no problem as it could draw on the national power grid. In space, power becomes a major problem. A nuclear reactor or a very large solar array are called for. A reactor this powerful for space use augurs political problems. Hostility to atomic power plants and the lingering memory of Cosmos 954 and 1402 are called to mind. The alternative of a very large solar array would be expensive and difficult to assemble in space, to say nothing of its vulnerability to attack and the risk of regular meteoroid damage to its large surface area.[16]

In either case, the reactor or solar cells would feed power into banks of capacitors, building up the power to a tremendous level. This accumulated energy is then discharged into the accelerator in a short burst, firing the beam. The build-up of power repeats itself while the battle station acquires the next target. Test systems use water capacitors in which pressurized water serves as a dielectric to store the energy; a sort of gigantic battery.

Size of the power source, and firing rate of the weapon are directly connected. The weapon designer has accordingly

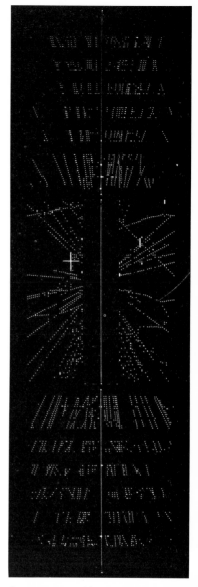

Computer graphic of proton-antiproton particle tracks in a so-called streamer (or bubble) chamber. These subatomic collisions are taking place in the gigantic CERN Super Proton Synchroton near Geneva, Switzerland. Other subatomic particles of quantum physics include mesons, hyperons, leptons and quarks (the last term borrowed from James Joyce).

to juggle the bulk of the power source and the weapon's firing rate. The smaller the power supply, the slower the firing rate. For ABM use, a weapon needs to fire every few seconds, whereas for ASAT and DSAT operations charge times of up to a few minutes might be acceptable.

A related design problem is insulating such high electrical currents. The slightest flaw in the insulation will result in an electrical discharge with spectacular results. A particular electron beam system used in manufacturing was fitted with an external nitrogen tank, the gas being supplied through a narrow-diameter, 30 ft (10 m) long Teflon pipe. When the machine was turned on, the current flowed through the gas and blew off a fitting on the tank. In spite of 30 ft (10 m) of nonconducting gas, the current still went to ground. This gives an idea of the problems encountered in constructing a full-scale weapon.

**The story
so far**
US particle beam weapons research is still in its infancy, perhaps ten years behind the work done on lasers. Particle beams are still confined to the laboratory; there has been no open-air testing yet, though this is planned. At the moment the power levels are still too low for weapons applications. There is nothing in this field of development comparable to the Navy's Sea Lite laser system. Indeed, nothing like the 1973 test at Kirtland AFB, New Mexico in which a target drone was shot down by a laser has yet taken place using particle beams. This is true at least of the US, where the various elements have not yet been combined in a total system.

Information about the Soviet particle beam effort began to emerge in the late 1970s. The picture that fragmentary reports paint is of a large-scale effort and considerable resources. The first Soviet test facility was built 35 miles (56 km) south of the town of Semipalatinsk at the southern edge of the Soviet nuclear test site. Its estimated cost is $500 million. The main building is 700 ft (213 m) long and 200 ft (61 m) wide with walls 10 ft (3.05 m) thick, and is thought to contain an accelerator, electron injectors and power-storage equipment.

Near the building four holes were sunk into solid granite and rock excavated to form large cavities deep underground. In another building nearby, large, extremely thick segments of a sphere 58 ft (18 m) in diameter were being manufactured. Each 12 ft (3.6 m) thick segment was transported to a shaft and lowered into the hole. Two complete spheres were produced in this way. The segments were apparently welded together by a process called flux welding, a technique used for several years in the USSR to produce heavy-pressure vessels. In this case, the joints are very likely as strong or stronger than the surrounding material. From satellite photos, Air

Force intelligence interprets the facility as a particle beam test bed.

According to their analysis, a small nuclear explosion set off inside the steel sphere is cushioned by liquid hydrogen. This explosion generates power for the accelerator, the energy being transferred through pressurized gas lines. Because of the huge power levels, these must be cooled with liquid hydrogen to prevent their burning out. (Ironically, such gas lines were originally developed by ITT and General Electric.) The power is then sent through transformers to step it up and stored in capacitors inside the large building. The Soviets were known to be using pressurized-water capacitors. With internal pressure equivalent to 100 atmospheres, it was calculated that they could store 40 times the energy of contemporary US systems.

Inside the large building is the accelerator used to produce a proton beam which is steered magnetically through kilometer-long underground drift tubes. The drift tubes were evacuated to test beam behavior in a vacuum with the idea of simulating what would happen in space. It is possible that the Soviets conducted open air testing at one time. Five rings of vertical sensors were placed around the large building spaced about 3.1 miles (5 km) apart. Each ring had 72 sensors positioned precisely 5° apart. At first they were believed to be for monitoring the hydrogen gas and radioactive nuclear debris vented from the sphere. It was however suggested later that the sensors were to measure beam effects and/or tracking.

Venting of large hydrogen clouds was first detected by US Air Force 647 early warning satellites in November 1975. A total of seven such clouds were detected between then and early 1977. The Air Force interpretation started a tremendous controversy within the US Intelligence community, pitting Major General George J Keegan against the CIA, a situation reflected in the designations given to the facility. The Air Force called it PNUT (Possible Nuclear Underground Test). The CIA referred to it neutrally as URDF-3 (Unidentified Research and Development Facility-3). The CIA Nuclear Intelligence Board believed it was impossible that the Soviets could have made breakthroughs undetected in no less than seven areas of technology. The necessary breakthroughs were: explosive power generation, capacitors, electron injectors, accelerators, flux compression to convert the energy, switching systems to store the energy and pressurized lines to transfer the energy.[17]

The arguments were reminiscent of the 'missile gap' controversy of the late 1950s when the Air Force believed the Soviets were deploying large numbers of SS-6 ICBMs while the CIA because of lack of evidence on U-2 photos had estimated that only limited deployment was underway. Reconnaissance satellites ultimately proved the CIA right.[18]

CIA analysts suggested various purposes for the PNUT/URDF-3 installation. Some 20 theories were put forward in all. They included a storage facility for unused nuclear fuel, a supersonic ramjet test site or an experimental nuclear reactor for commercial operations. The latter was suggested because the layout resembled that used in Soviet reactors. General Keegan refused to accept the CIA interpretation. He assembled a team of young physicists to examine each of the seven technological stumbling blocks. They were able to show to their own satisfaction at least that the Soviets were able to build the PNUT facility. The evidence included data from reconnaissance and early warning satellites, as well as Soviet scientific papers and contacts between US and Soviet scientists. General Keegan ultimately retired in protest at the CIA's unwillingness to accept his view of a large-scale Soviet particle beam weapons effort.[19]

The Soviets subsequently made ground tests of the effects of particle beams on re-entry vehicles, high explosives and other materials. These ground tests were read as preparations for open-air testing.[20]

The Saryshagan enigma

The Air Force and General Keegan were vindicated by the discovery of a new facility codenamed Tora. (Tora is the Japanese word for tiger, and was also the Japanese codeword for the attack on Pearl Harbor.)[21]. Work had begun in November 1979 at the Saryshagan ABM test site. The installation was long and slender; its configuration resembles that of a linear accelerator with the particle source at one end and the beam-aiming system at the other. It was noted that Tora was equipped with a movable nozzle. This nozzle was interpreted as a device to allow tests against moving targets such as aircraft or missile warheads. The accelerator was connected by a series of cables to a building which presumably housed power-switching equipment. Another set of cables ran from this building to a row of 12 magneto-explosive generators. Also called Pavlovski generators, they are able to produce a brief, intense pulse of electrical energy for the weapon.[22]

A conventional high explosive is used to compress a magnetic field, generating the electrical current. The Pavlovski generators replace both the power source and the bank of capacitors. Rather than taking the energy from a source and feeding it into the storage capacitors to build up the charge, it is fed directly from the generators. These generators are 'single shot' and before they can be fired again have to be 're-loaded'.[23] Because of its external appearance and the use of Pavlovski generators, Tora was initially thought to be an electron beam weapon. Others guessed that it was a pulse-iodine laser after a suggestion that the Soviets were exploring this technique. Such a device, however, was expected to be short and compact.

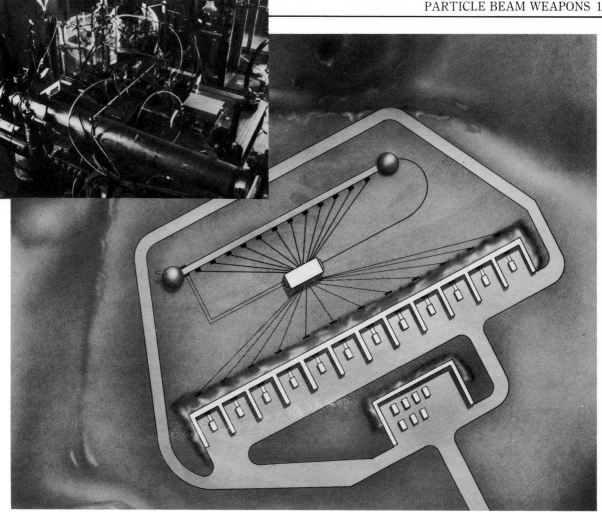

The US had high-resolution photos of the system's exterior but the inside was still unknown. Subsequent investigations indicated that Tora was an iodine laser despite its appearance.

Tora had conducted about 30 tests against re-entering warheads by early 1981, some of which were successful. To assess the particular advantages, limitations and technology of Tora, the Air Force awarded a contract to the Los Alamos Scientific Laboratory to build a small-scale iodine laser. The Los Alamos device was so successful that the Air Force afterwards sought funding to build a larger version for testing against real targets.[24]

Artist's impression of the beam weapon at Saryshagan, from the air. Situated near the Sino-Soviet border and codenamed Tora by US observers, it is a hybrid weapon combining laser and particle beam features. It resembles the iodine laser (inset), but on a much larger scale. Note the 12 explosive generators which provide power, and the moveable 'nozzle' which allows tests against moving targets – such as re-entering warheads. General George J Keegan argued that a similar facility is presently installed at Semipalatinsk, and was eventually vindicated in this.

Lasers, 'wigglers' and X-rays

Tora is an example of a third weapons option, namely combining laser and particle beams to produce a hybrid. The result is a laser which uses explosive generators for power, devices normally associated with particle beam weapons. This explains the confused speculation about Tora. Another example is a Soviet ASAT laser based near Moscow, which appears to be a carbon dioxide gas-dynamic laser but is pumped with an electron beam.[25]

Several types of hybrid are proposed or actually in development. One is the Excimer laser. The name comes from the chemical term 'dimer', meaning a compound formed from two molecules of simpler compounds. The laser uses excited dimer and so is known as an Excimer laser. First proposed by scientists at Avco Everett Research Laboratory, work has concentrated on gas mixtures of krypton-fluorine and xenon-fluorine. (These two gases do not normally react. Krypton and xenon are inert gases but when krypton is excited it chemically resembles rubidium and can then combine with the fluorine. However, such a combination is not stable; energy is lost in the form of a photon. The compound is then no longer excited and the two atoms separate.)[26]

The 'pumping' is done by an electron beam, so ionizing the gases as to create the necessary conditions for the Excimer effect. Output is in the ultraviolet wavelengths.[27] This shorter wavelength is advantageous since it means that the segmented mirror can be smaller (a visible-light laser needs a large mirror by comparison). The Excimer laser is also efficient, with ten or 20 per cent of the input energy being converted to laser energy. Improvement to a level of perhaps 50 per cent is looked for. This would be a very considerable achievement compared with present typical efficiencies.

A related system is the blue-green laser whereby the Excimer's ultraviolet light is sent through lead vapor or hydrogen gas which converts it into coherent blue-green light. The specific color of a laser depends on the electron configuration of the lasing material which is employed. This wavelength has the ability to penetrate air and water to a considerable depth, making it useful for communicating with submarines. The blue-green wavelength can be achieved directly by a Mercury bromide or thulium-YLF laser.[28] Use of a blue-green laser for submarine communications has been tested using a laser installed in a T-39 aircraft, the receiver being aboard the research submarine USS *Dolphin*. Results were evidently impressive.

This technological advance in undersea communications has important implications. Submarines currently have to come close to the surface to receive radio orders, a very vulnerable position in wartime. A blue-green laser would allow them to receive orders while deeply submerged and so less likely to be detected.

Another blue-green laser test program concerns the Air Force's LaserCom carried aboard an EC-135 (though not the ALL aircraft). Ground tests were made in September 1978 at the White Sands missile range, in which a 1.24 miles (2 km) long optical path simulated the worst air-to-ground conditions imaginable; the test indicated that the system would work. The next step was two-way communications between an aircraft and ground station, as tested at White Sands between October 1979 and December 1980. To test this the LaserCom EC-135 flew in circles around the ground station at altitudes between 33 000 and 37 000 ft (10 058 to 11 278 m). The distance separating the aircraft and ground station was up to 62 miles (100 km). The ground station and aircraft were able to acquire each other and could 'talk' back and forth.

In the final test, live video and a videotape were transmitted by laser from the EC-135. These transmissions were received by the ground station, then videotaped, and returned to the aircraft by laser. The signals were finally re-transmitted to the ground station. When the transmissions were compared with the original tape, there was no visible difference between them.

The first space test of laser communications is to involve the Teal Ruby satellite, which will be in a 400-nautical mile (741-km) orbit, inclined 72.5°. A small laser package aboard the satellite will test acquisition, pointing and tracking of the White Sands ground station during its two 10-min passes each day. The effects of the atmosphere and the system's ability to handle multiple users will be investigated.[29]

Two operational techniques for the blue-green laser have been suggested. One is to put the laser into space, trans-mitting the messages up to the laser-satellite by radio for re-transmission by the laser, an approach which calls for a high degree of laser reliability. The other is to keep the laser on the ground and use instead a large orbiting mirror to relay messages to the submarines, an approach which calls for a much more powerful laser and which might be hindered by atmospheric problems.[30]

Fast talk from the laser beam

A laser communication system has several advantages over conventional radio. First and foremost, the laser has a shorter wave-length than radio frequencies (0.00053 mm as opposed to between 1 mm and 1000 km). This means that a laser promises an extremely high data transmission rate: in theory up to one billion bits per second, equivalent to trans-mitting the entire text of the *Encyclopedia Britannica* in a single second. What is more, because a laser is used, the area covered by the signal would be extremely small. This is in marked contrast to even the narrowest radio signal from a geosynchronous orbit satellite, which is about 200 miles (322 km) in diameter. A laser communications beam from the same altitude would only be 0.1 mile (0.16 km) wide.

An enemy would be hard pushed to eavesdrop on such a narrow source. Similarly, it is difficult to locate such a satellite by homing in on its signals. It would be false however to claim that this narrow beam is without danger for submarines. If an enemy could somehow discover just where the beam was pointed, he would know the exact location of the submarine. One answer to this which immediately suggests itself is to lay false trails by beaming the laser at a great number of points within the ocean; so many that an enemy's anti-submarine forces could not cover them all.

It is a telling detail of the current international situation that laser communications are subordinated to laser weapons research. The EC-135 laser communications test system is 200 milliwatts; nine orders of magnitude less than a weapon's requirement. Laser telephones come second to laser battle stations.[31]

Yet another hybrid along these lines is the Free Electron Laser, an offshoot of research on particle accelerators. In this technique, an electron beam is sent through an alternating magnetic field which varies in intensity and direction. Because of its similarity to an undulating worm, the beam is called a 'wiggler'. As it travels through the magnetic field, the beam gives off infrared energy. Mirrors are placed at opposite ends of the optical cavity so that light reflects back and forth and becomes a laser beam.

The Free Electron Laser uses only electricity; there is no lasing material and no large gas supply is required. Another peculiarity is that the wavelength can be adjusted to a certain extent and can compensate in this way for atmospheric conditions.[32,33] Finally, the Free Electron Laser offers high efficiency since 30 per cent or more of the energy input is converted to laser energy.

The principle was first suggested in 1951 by a Stanford University scientist as a monitoring procedure for particle accelerators. Another Stanford scientist, John M J Madey, proposed 20 years later that a laser could be developed using this principle. Power levels of several kilowatts had been attained by the spring of 1977. Interest in Free Electron Lasers centers on ground-based use, as this type of laser has the potential to adjust to atmospheric effects. Nor would electrical power, size or weight be problems.

By far the most unnerving member of the hybrid family is the X-ray laser. No normal laser design can use X-rays. Even if the lasing material were somehow made to emit X-rays, they would go right through the mirrors. The X-ray lasers under study employ only a single pass through the lasing material.

Two designs have been proposed: the X-ray pumped laser and the pure X-ray laser. Both use a small atomic bomb as the energy source. The first has at its center a 40-kiloton nuclear bomb; surrounding it would be 50 separate laser units. Containing krypton fluoride gas, the laser employs the Excimer laser technique.

In one scheme, each laser is individually pointed at its target by rockets with an accuracy of 0.5 microradians. The lasers would be 6.6 ft (2 m) in diameter and the complete ring 853 ft (260 m) across, with lasers spaced 52 ft (16 m) apart in a configuration resembling a huge spiny sea urchin.

A low-level laser acquires and holds each target; the krypton fluoride gas is then emitted and becomes uniform. Once all the lasers are tracking their targets, the nuclear warhead is detonated, and the X-rays emitted are converted into high-energy electrons which pump the krypton fluoride gas just like an electron beam. All 50 lasers would then fire a single brief pulse; and 1/1000th of a second after the nuclear weapon is detonated, the expanding fireball consumes the laser itself. By that time, however, the beams would be headed towards their targets.[34]

A possible alternative, assuming that difficult problems are eventually solved, is the pure X-ray laser. It uses a similar configuration but different materials. The nuclear warhead is surrounded by short, small-diameter metal rods with a high atomic number, only 3 to 8 ft (0.9 to 2.4 m) long. The complete package would be small enough to allow a single Shuttle launching to put several into orbit (in contrast to the several launches needed to construct even a single laser or beam weapon battle station). The X-ray beams generated by this system would have an intensity of several hundred trillion watts. The shock wave upon impact would travel through a target's structure and emerge from the other side with catastrophic effects. The target would be irradiated, incinerated and fragmented simultaneously in a 'kill time' of no more than 1/1 000 000th of a second.

X-ray lasers could be permanently stationed in space or kept on the ground, ready for launch in a crisis. They could serve as long-range ABMs, launched when an attack was underway. After going into orbit, they would destroy enemy ICBMs before they are consumed by the fireballs they generate themselves. They seem to be the last word in sophisticated, self-destructive weaponry.

An end to MAD?
In addition to the technological questions, beam weapons face major political problems. These may be more formidable than any technical ones. Such weapons are politically inconvenient because they re-open questions of strategy and nuclear doctrine thought long settled. Specifically, they challenge the doctrine of Mutual Assured Destruction. For two decades, this has been the cornerstone of US nuclear policy and arms control efforts. Simply put, MAD holds that no matter how the Soviets attack, enough US missiles, bombers and submarines will survive to destroy the USSR. In the event of a Soviet first strike, the US would exact a posthumous revenge. No reason, it is argued, could be worth the millions of deaths, scorched cities, and other awful consequences of such a nuclear

exchange, and so both sides hang fire.[35] MAD provides defense by threat of offense. Accordingly, it lays emphasis on the efficiency and quantities of offensive weapons.

Purely defensive measures are low-priority. Defenses such as ABMs for cities or industrial centers are condemned as de-stabilizing factors. Cities are hostages under MAD and the threat of their destruction is intended to prevent nuclear exchange. Measures such as ABMs, shelters and civil defense are de-stabilizing because, if they worked, a society could survive an attack (or at least give this illusion). Under these circumstances, the reasoning runs, a nation might be more inclined to use nuclear weapons. In peacetime, an ABM system tends to create pressure for bigger, more powerful offensive systems to punch through the defense, thereby creating another spiral in the arms race.[36] These arguments were marshalled against the Nike-Zeus ABM and apply even more strongly to orbital beam weapons.

Such weapons promise to destroy attacking missiles before they ever reach their targets. Under MAD, this is inherently de-stabilizing. The doctrine of MAD is jeopardized by technological developments like these, which upset the nuclear relationship between the US and the USSR, a relationship correctly described as a delicate balance of terror. Consider the similar example that development of a super-accurate missile guidance system provides.

ICBMs, in their fixed silos, may in this case be destroyed with a high degree of confidence, opening up the prospect of a disarming first strike. This removes the 'mutual' and leaves only 'assured destruction' for one side. The deterrent which MAD represents is fundamentally a matter of belief, since it hinges on fear of a counterstrike. MAD has led to the current predicament in which missiles are protected while the population is left defenseless.

The argument runs that under MAD opposed nuclear forces are counterbalanced, so that a nuclear attack is no longer a rational option; there is only the minimal risk of an exchange through insanity or an accident.[37] MAD holds that nuclear weapons are not like conventional weapons such as ships, airplanes and tanks: the destructiveness of nuclear weapons means that there is no possibility of winning a nuclear war.

Their power also renders 'superiority' meaningless. When both sides have thousands of warheads, a few more makes a negligible difference. Finally, any use of nuclear weapons would inevitably lead to a final convulsive spasm in which all weapons are fired.[38] This nuclear war, in which there is no victor and which cannot be survived, is supposedly unthinkable. 'We are all dead' was how one administration spokesman described the situation after a projected nuclear exchange.[39]

MAD was formulated during the early 1960s when it

made technical sense. The early ICBMs were inaccurate and cities were the only things they could hit; they were unlikely to destroy hardened missile silos. The early Atlas and Titan ICBMs needed a long time to fuel and prepare and so could only be used in an all-out attack.[40,41] Antimissile defenses at this time were ineffective and extremely expensive. The Nike-Zeus radar, for instance, could not separate warheads and decoys, nor cope with a saturation attack, not to mention the fact that it was vulnerable to low-altitude nuclear blasts.[42]

Time and technology have left MAD behind however, in the view of its critics. Guidance systems have now improved to the point where fixed missile silos are no longer safe. The 'window of vulnerability' question focuses on the possibility that the Soviets could destroy US Minuteman missiles with a small part of their SS-18 forces, holding the rest in reserve to enforce terms of surrender.

Critics also contend that MAD exaggerates the effects of nuclear explosions and does not take into account the differences between an attack aimed at cities and one directed against weapons.[43] In their view, it also fails to take into account the possibility of a limited nuclear attack aimed at a few selected targets as a 'demonstration'. These claims are met by the charge that talking about a 'limited' nuclear war is dangerous because it brings the possibility closer. According to this argument, the whole notion of a controlled nuclear exchange is unreal because the situation would inescapably and rapidly become an all-out exchange.

Claims that the effects of a nuclear war have been over-estimated are met by counterclaims that full-scale nuclear war would destroy virtually all life on Earth.

MAD has also been attacked as a morally indefensible doctrine, an argument which has gained importance with the resurgent Ban the Bomb movement and the US bishops' pastoral letter on nuclear war. MAD threatens to kill hundreds of millions of innocent people; even to prevent war this threat is morally repugnant.[44] Under MAD, a retaliatory strike is a vengeful act of mass murder. According to MAD, if an exchange takes place it inevitably leads to total destruction. MAD's dogma that numbers do not matter, along with a vocal anti-nuclear movement, makes it increasingly difficult to maintain a deterrent.

From the mid-1960s, and throughout the 1970s, US nuclear forces remained frozen in terms of the total number of missiles to deliver warheads, although warheads themselves continued to be manufactured. With the retirement of early model B-52s and the Titan II force in the early 1980s, this too diminished. Like most American institutions, MAD has suffered a battering over the last few years.

Beam weapon advocates insist that there is a third option between annihilation and surrender. Beam weapons, they argue, offer hope that the competitive cycle of ever more powerful offensive systems may be broken and the shadow of MAD lifted.

Beam weapons — for and against

Beam weapons have naturally given rise to heated controversy. Broadly speaking, the arguments against them fall into three categories. The first might be called the scientific intellectual view, and emphasizes the formidable technical problems which constructing a beam weapon poses. The second is the arms control position which sees beam weapons as the joker in the pack of international arms agreements. And the third is the military view which considers beam weapons useless because very vulnerable.

To treat the respective positions in further detail, the scientific intellectual argument points out the great degree of accuracy required by a beam weapon and the difficulties involved in achieving this. Such achievement is considered to be a long way in the future and to demand much more attention before final commitment to a large program.[45] The argument parallels those against early US ABM programs.

One difference between the earlier arguments about armaments and the present ones is the rise of the military reform movement. This group maintains that numbers alone are important and that technical sophistication counts for nothing. According to this viewpoint, it is preferable to build large numbers of (for example) austere fighter aircraft such as the F-5 or F-20 rather than the more complex (and so less readily maintainable) F-14 or F-15s. The same holds for other military equipment: small carriers rather than super-carriers; diesel submarines rather than nuclear ones.

They insist that modern technology is unreliable, costs too much, is accident-prone and hard to repair: the US military is seen as infatuated with ever more complex technology for its own sake, producing weapons which are less effective in real combat situations.[46] Others hold that on the contrary *simple* systems are ineffective in a modern combat environment. There is, for example, no such thing as a simple all-weather fighter, since an all-weather fighter must have complex long-range radar and missiles of the sort that the military reformers criticize.[47] Modern technology (they argue) is more reliable, cheaper and of higher quality than previous systems (TV sets, calculators and arcade video are cited as examples). This school of thought locates the real problems in the *age* of the equipment (as much as 20 years old in some cases), lack of spare parts and loss of trained personnel to higher paying civilian jobs. These factors make simplicity or complexity of a system quite irrelevant, since the odds are stacked impossibly against it from the start.[48]

The scientific intellectual view and the military reform movement pertain very directly to beam weapons, the former arguing that the technology to build them does not exist and the latter that even if they can be built, they will constitute a wholly unreliable (because highly sophisticated) system. [49]

The military reformers quickly point out as well that beam weapons — especially space-based ones — make tempting and vulnerable targets.[50] Future debates over beam weapons are perhaps prefigured by the conflicting interpretations of the Falklands War. From the same events, it is possible to 'prove' that super-carriers were or were not obsolete.[51]

The extreme arms control position is simply that beam weapons must not be built. They are perceived as having a destabilizing effect on the global balance of power. In this respect, MAD and arms control have similar philosophies. MAD assumes that an uncontrolled arms race between the US and the USSR lessens each side's security rather than increasing it by making a nuclear exchange more likely. On the other hand a mutual deterrent with each side capable of destroying the other is intended to create a better political climate.[52]

Approached from this angle, the US only threatens itself by deploying beam weapons. The Soviets are then compelled to develop a counter, thus accelerating the arms race. If the US does nothing, the arms control argument runs, then neither will the USSR.

The arms control prescription for removing the Soviet space threat is for the US and USSR to negotiate a treaty banning all weapons in space. This covers not only orbital nuclear bombs but ASATs, DSATs and beam weapons. The UN might act as a forum in which every nation involved is called upon to declare and describe all its space activities. This removes the threat of a costly and futile expansion of the arms race.[53] Such was the underlying philosophy of the ASAT talks instigated by President Carter. US technical superiority was to be traded against the operational Soviet system, with a ban on space weapons as the goal.

Another argument against beam weapons is that they might bring about the very exchange they are meant to prevent. It has been suggested that if the US began to launch ABM battle stations, the Soviets may interpret this as preparation for an attack. The Soviets, while they still had the chance, would then launch a pre-emptive strike.[54]

It is interesting to note in this connection the events of August and September 1969, when after clashes between Soviet and Chinese forces the Soviets began to circulate reports that they were contemplating a nuclear attack on the Chinese. This went so far as 'unofficial' enquiries about the US reaction to a Soviet pre-emptive strike. It may be that these reports were meant to pressure the Chinese into resuming the border talks, so that the threat itself was largely a bluff. This threat was perhaps one reason for the US-Chinese talks that led to President Nixon's visit and ultimately to diplomatic recognition for China.[55] The Chinese had at the time only a small number of nuclear weapons, and their delivery systems were limited to Soviet-built copies of the USAF's B-29 bomber (TU-4s), supplied to China in the mid-1950s and easily dealt with by Soviet air defenses.[56]

The final set of arguments — representing the military view — is the opposite of the arms control position. Defense Secretary Harold Brown's remark encapsulates the US military assessment of the current situation: 'we build, they build: we reduce, they build'.[57] Ironically, the arms control and military positions are both opposed to beam weapons. The military is committed to defense through offense, arguing that the offensive stance always has the advantage, and stressing the vulnerability of orbiting battle stations. Very cheap counter-measures could defeat the battle stations, which are depicted as being bulky and even fragile.

Remember the Maginot Line An example quoted to illustrate this possibility is the Maginot Line. The pride of France in the 1930s, it was meant to be an impenetrable barrier to German attack, equipped as it was with minefields, heavy barbed wire and anti-tank traps. The Maginot Line was quite impenetrable to a direct attack, it was argued. Germany would bleed itself white against these fixed fortifications — assuming of course Germany attacked on its terms. But there was a loophole: the Maginot Line's northern end stopped at the French-Belgium border; the Ardennes, with its hills and forests, were deemed unsuitable for tanks. It was here that the offense-minded Germans attacked. A well-contrived offensive plan rendered it useless and the Maginot Line has been a byword for failure ever since.[58,59]

Another concern of the military viewpoint is the effect a commitment to a large-scale beam weapons program would have on existing strategic development programs. The argument goes that beam weapons are still many years, if not decades, away from serious application. Advantages derived from such a program now would be largely limited to propaganda.

Perhaps the most serious obstacle to beam weapons is formed by existing arms control agreements. The 1972 ABM treaty and later amendments prohibits development or testing of sea, air, space or mobile land-based ABM systems. A fixed-site land-based beam weapon might be developed and tested, taking advantage of a loophole in this treaty. Beam weapons, however, could be considered to be based on new physical principles; in which case the treaty specifies that limitations would have to be determined through discussions between the US and Soviets.[60] For this purpose, reviews are scheduled every five years, the next two in 1987, and 1992, by which time beam weapons may be more feasible.[61]

X-ray lasers pose a notable problem in this regard. Incorporating a nuclear bomb, they stand to violate the 1967 space treaty. Although not a surface bombardment system like the FOBS, they constitute orbital nuclear weapons. This might be circumvented by building them anyway but launching them only in a crisis or when an attack was underway. Testing an X-ray laser, after all, can probably be done underground, outside the nuclear test ban treaty's jurisdiction.

Beam weapons are fundamentally different from any previous type of weapon. Their development is not so much a matter of money — it costs tens of billions of dollars to design, develop, produce and operate a new jet fighter. The kind of arguments commonly levelled against tanks and aircraft — conventional weaponry — are also levelled against beam weapons (both involve high technical risks and cheap countermeasures). It is rare, however, that these are a major impediment in the case of tanks and aircraft.

Already this accumulation of objections is taking its toll on beam weapons. In 1980, the US development effort was changed from near-term Earth-based lasers, placing the emphasis instead on space lasers. In the spring of 1982, the House Armed Services Committee recommended cancellation of the Alpha 2 to 3 megawatt laser, the Large Optics Demonstration Experiment program and a $40.6 million request for an Air Force study of space lasers. Emphasis was to be shifted from chemical lasers to shorter wavelength systems such as the X-ray, Free Electron and Excimer laser types. The House Committee recommended adding $50 million to accelerate short wavelength laser research.

Another suggestion was that the Navy's Sea Lite laser be switched from investigating sea-based air and missile defense to fulfill a ground-based ASAT role. The recommendation generated controversy. Many argued that the short wavelength lasers would not match existing chemical lasers for many years. This would put off even an experimental space laser until the mid-1990s. Far from accelerating the US laser program, these recommendations would cripple it. The changes were eventually rejected but the time spent on them and the different directions such projects can take are amply illustrated by this example.

Beam weapons in 2015

There are other important factors to consider. For instance, the Department of Defense tends to study problems endlessly but is reluctant to innovate. A major strategic system may take 20 years or more to progress from concept to operation. Studies of the B-1 strategic bomber began in the early 1960s. Serious development began in the early 1970s. The B-1 then underwent a roller coaster ride of flight tests, cancellation, revival and modification. The B-1 will still not be operational until the late 1980s at the earliest — a quarter of a century after its conception.

The concept of a ten-warhead super-accurate ICBM first emerged in the US and Soviet Union during the mid-1960s. The Soviet version — the SS-18 — became fully operational in 1980. Its US counterpart — the MX — has been dogged by political controversy. If there are no further political objections, it will be fully operational in the late 1980s or early 1990s.[62,63]

Even if an ABM beam weapon system is demonstrated to be feasible, the B-1 and MX experiences indicate that it could be the year 2015 or 2020 before such an ABM network is operational. The delay would not be wholly due to technical problems, but also to the inefficiencies of the management process. In reply to these arguments, supporters of beam weapons stress that they promise escape from the mutual hostage relationship of the US and Soviet Union.

This shift to a defensive posture would be gradual. Beam ABMs have only a limited capability against a mass ICBM attack: their role is damage limiting. Traditional offensive spending remains important alongside them. The race would then be for increasingly powerful defensive weapons. In this way, the transition could perhaps be accomplished successfully with a minimum of danger.[64]

Beam weapons supporters believe essentially that they have the answer to questions that MAD does not address. Just how mutual for example is 'mutual deterrent'? Implicit in MAD is the questionable belief that both the US and the Soviets accept its doctrine and consequences. The Soviets have deployed an ABM system around Moscow and are undertaking a civil defense program to protect the political leadership, essential work force and industrial capability. Does this suggest that the Soviets may not consider a nuclear strike 'unthinkable', given desperate circumstances?

Another blind spot in MAD which beam weapons remove is the assumption that nuclear deterrence will always succeed; that rationality will always prevent an exchange. Murder of whole segments of a population is not unknown in the modern world and is even considered in some ideologies to be an acceptable part of exercising political power. Not only Nazi Germany, but Uganda and Cambodia spring to mind. And as Winston Churchill noted in 1955, deterrence does not cover Hitler in his bunker. Beam weapons, supporters argue, offer the promise of removing the threat of nuclear annihilation by making nuclear weapons, in the very long term, obsolete. Missiles could not penetrate the defense in sufficient numbers to annihilate the adversary. In this way, the possibility of a disarming first strike would be mitigated. Whatever argument one subscribes to, it seems that by the end of the 1990s, the issue will be decided.

Chapter 7

CONCLUSION

'It is a tragic paradox of our age that the highly humane objective of preventing nuclear war is served by a military doctrine and engines of destruction whose very purpose is to inflict genocide.'

Fred Charles Ikle
Every War Must End (1971)

The battle for space begins in the laboratory, in the corridors of power and, most importantly, in the minds of men. Mankind has only been able to launch objects into orbit for little more than a quarter-century and laws to govern space activity are in the embryonic stage. Because the battle for space is an extension of political rivalry on Earth, similarities exist between it and earlier military strategies. Earth-based strategies for instance place great emphasis on air superiority and control of the sea. In space, the equivalent is perhaps the attempt by both protagonists to operate satellites and make full use of the information thus supplied while hindering and checking the enemy. The loss of a reconnaissance satellite under these circumstances is roughly equivalent to losing an SR-71, U-2 or MiG-25 aircraft in conventional warfare. Destruction of a communications satellite disrupts the relay of orders as effectively as bombing a radio station.

A more subtle parallel is offered by the age-old admonition to 'take the high ground' — that is, to occupy a commanding position on a hill or raised terrain. Space is effectively the high ground for the entire Earth.

If space war breaks out

What form would the battle for space actually take if a US-Soviet war broke out? Let us consider first the near future (the second half of the 1980s) and then the longer term (the early years of the 21st century).

By the mid or late 1980s, both the US and the Soviets will have operational ASATs. In the US camp, two ASAT F-15 squadrons will be trained and in place alongside the necessary interceptor missiles, the pilots and ground crews still new to their role. The Soviet ASAT launch crews, on the other hand, will have two decades of experience behind them.

At the outbreak of world war, the Soviets would attempt to destroy US reconnaissance and communications satellites, initiating at the same time a ballistic attack on land targets. As the lights on the NORAD status board winked out, signifying the destruction of each satellite, the Soviet armored thrust would be crossing Western borders. The Western Allies would strike back in space as on Earth.

The USAF's F-15 squadrons would try to knock out the Soviet ocean surveillance satellites, clearing the way for the safer deployment of shipping convoys, an effort matching the Navy's attempts to hunt down Soviet submarines and surface ships. Meanwhile the F-15 ASATs would be inflicting as much damage as possible upon Soviet space systems. At the launch sites, exhausted crews are working to put up replacement satellites, each side nervously eyeing the other's activity with a view to attacking launch pads and disabling satellites on the ground. Display screens in NORAD's command post are furiously busy.

Although both sides by this time have ground-based lasers powerful enough to disable low-orbit satellites, they would still largely be in the teething stage of development. In the 21st century however, lasers are expected to play a central role. Although the goals would be much the same, targets, tactics and weapons may differ. Permanently manned stations, orbital factories and solar power stations are all potential future targets. (By this time, moreover, space may no longer be the exclusive domain of the US and the Soviets.) Ground-based lasers capable of destroying satellites in orbit will very likely be fully operational, replacing conventional ASAT systems (such as the F-15 or the orbital interceptor). These lasers are anticipated to be powerful enough to destroy satellites in low orbit, and perhaps those at geosynchronous altitudes as well. Orbital lasers are expected to be available, ranging from simple ASATs to a sophisticated damage-limiting ABM network.

These will alter the form of the conflict considerably. Conventional ASATs prey on satellites; because of timing and orbital mechanics, however, it is extremely difficult for one predatory ASAT to destroy another. With a laser's ability to strike at long range, this no longer will be the case and the satellite predators become the prey. A laser gunfight may take place in space. As the lasers attack other satellites, they are sniped at by defending lasers. Down below on Earth's green acres, the final outcome is anxiously awaited.

Such weapons will seriously affect international relations and nuclear strategy whether or not they are fired in anger. They are the latest chapter in a story as old as organized warfare.

It is predicted that *early* ABM lasers will be capable of mitigating damage but unable to stop an attack outright. They may protect ICBMs in their silos from a disarming first strike, the role envisioned for traditional ABM missiles, but the composition of strategic nuclear forces is not expected to change radically in the short term. ICBMs are not about to vanish. Indeed, ICBMs may be polished or coated with ablative materials as protection against laser beams. The only real change is likely to be that a percentage of ICBMs — perhaps up to 30 per cent of the strike force — are disabled or detonated before reaching their targets. To military thinking, a percentage as low as this is not sufficient reason to abandon the ICBM outright. After all, both the US and USSR fly bombers, even though they know for a fact that many stand to be lost in the event of conflict. However, because the ICBM is no longer invulnerable, the notion that it is now obsolete may gather momentum.

In the mid-term, as the orbital network ramifies and the battle stations become more powerful, the position of the ICBM will certainly be undermined. Encouraging as it sounds, this does not in itself necessarily spell an end to nuclear weapons, since if the US and USSR want to maintain an offensive nuclear capability, bombers and cruise missiles may remain as the primary means of attack. Flying at low altitude, and concealed by the atmosphere and cloud cover, they are out of range of the orbiting lasers. Stealth aircraft are becoming increasingly sophisticated and — having low radar response and infrared guidance systems — are much more difficult to detect than an ICBM. (Although a missile might be covered with radar-absorbing material, little can be done about the mile-long and highly visible exhaust plume.)

A ray of hope? A major and desirable change in nuclear strategy may result, all the same. ICBMs carry with them the potential of instantaneous destruction and are so fast that the size of a country is an unimportant factor — a missile aimed at North Dakota would strike only a little earlier than one aimed at Texas. Bombers, even at supersonic speed, take time to reach their targets, so that a bomber attack spans an hour or more. This allows an adversary's bombers to scramble once an attack is underway and so precludes a disarming first strike.

Most importantly, and again in marked contrast to ICBMs, beam weapons can be so manipulated as to destroy only a narrowly defined area — a building rather than a city. This removes at a stroke the terrifyingly indiscriminate destruction of an ICBM with its nuclear 'kill radius'. This promised reduction in the scale of violence is perhaps the most attractive feature of beam weapons and accounts for the widespread public interest in them. Innocents, it seems, may not have to live indefinitely in fear simply because their homes are within miles of a target.

Projections aside, a few things are certain. First and foremost, beam weapons are real. Although many years may elapse before they are operational, they are discernible on the horizon. To simply ignore them because they seem speculative is the most perilous course of all.

The arguments for and against beam weapons *must* be addressed. The choice is a major one involving not only great sums of money but crucial issues of strategy and foreign relations for the next 30 years or more. The airplane, tank, submarine, mass aerial bombing, nuclear weapons and the ICBM were all greeted with scepticism, yet each altered the nature of warfare. The arguments against beam weapons tend to assume that US and Soviet strategic weapons will remain unchanged for the next 30 or 50 years. This is simply not so, and if beam weapons can call a halt to the offensive weapons spiral, then they deserve a serious hearing.

Beam weapons belong to the world as it is rather than as it should be. Disarmament seems as elusive and distant a goal today as at the end of World War II. Force still tends to prevail in international disputes and ours can hardly be called in this respect an age of enlightenment. If beam weapons are an authentic alternative to the current state of affairs and a prospective peace-keeping system, it is right to explore them.

But projections are vain beyond a certain point and visions of tomorrow are finally a question of individual temperament. In the last analysis, nothing can safely be said about the future beyond the observation that it will be at least as surprising as the past. The hope that lasers and particle beams are a practical way out of our current dilemma may prove illusory, however much mankind wants to believe it. As a remedy for over optimism, it is as well to remember that no defense system yet devised has been anywhere near 100 per cent effective. Will ray weapons buy time for wiser counsels to prevail in world affairs? The future alone will tell.

Glossary

Ablative material A temperature-resistant material coating heat shields which slowly melts and reduces the heat of re-entry.
ABM Anti-ballistic missile. A rocket used to destroy a ballistic missile in flight.
Achromatic lens elements A set of two or more lenses which bring all wavelengths of light to a single focus.
Acquiring The act of detecting and tracking an object such as a satellite.
Acquisition and tracking radar Radar equipment which searches for, acquires and tracks an object by means of reflected radio-frequency energy from the object, or by means of a radio-frequency signal emitted by the object.
Actuator A device which acts on electronic signals (eg, from the guidance system), converting these into mechanical movement.
Altitude The distance in degrees from the horizon to an object above.
Apogee The high point of a satellite's orbit. The point at which it is furthest from Earth.
Atom bomb A bomb that derives its power from splitting atoms of uranium or plutonium.
Attitude The position of an aerospace vehicle as determined by the inclination of its axes to some frame of reference.
Azimuth The distance in degrees from true north along the horizon to an object.

Ballistic trajectory The curved portion of a vehicle trajectory traced after the propulsion force is cut off.
Bit A unit of information carried by an identifiable character, which can exist in either of two states – a *one* or a *zero*. An abbreviation of binary digit.
Booster An engine that assists the normal propulsive system of a launch vehicle.

C band A radio frequency band of 3.9 to 6.2 gigahertz per second.
Chaff Strips of aluminum foil cut to a particular length in order to reflect radar signals. When a large amount of chaff is released, a radar will register thousands of 'targets'.
Cryogenics Methods of producing very low temperatures.
Cryogenic gas A gas which has become so cold it is turned into a liquid.
Closed loop A control system which is capable of self-correction.

De-orbit The act of causing a satellite to leave orbit and return to Earth.
Dielectrically heated Heating while producing power (ie, the fuel cell).
Doppler effect The change in the frequency of a sound or radio signal caused by the movement of the source. As the source approaches the stationary listener the frequency is higher. As it passes, the frequency is at its normal value. As the source moves away the frequency is lower. This was first discovered by the German mathematician Christian Doppler.

Earth resources package A set of cameras aboard a satellite, used to monitor, among other things, the condition of crops on Earth as well as air and water pollution.
Eccentric Of an orbit, deviating from the line of a circle so as to form an ellipse.
Electromagnetic pulse (EMP) An extremely powerful current generated by the interaction of the fireball of a nuclear weapon exploding in space, and the Earth's magnetic field.
Elliptical orbit An orbit of which the high and low points differ by a significant margin.

Geosynchronous or Clarke orbit A 22 300-mile (35 880-km) high circular orbit. A satellite in such an orbit takes 24 hours to complete one revolution. It therefore seems to remain fixed over one point on Earth. The orbit's usefulness for communications was first noted by the science fiction writer Arthur C Clarke in the October 1945 issue of the magazine *Wireless World*.
G force The force on an object caused by the acceleration of a rocket in flight. This force is calculated in terms of the Earth's gravity. One G is the measure of the gravitational pull required to move a body at the rate of about 32.16 ft (9.8 m) per second per second.
Gimbal A framework that permits an object to move in two ways.
Ground track The theoretical mark traced on the surface of the Earth by a flying object or satellite as it passes over the surface.
Gyro system A set of spinning wheels (gyroscopes) that measure the changes of direction and speed of a rocket or satellite. The data is then sent to the guidance system so corrections can be made.

High orbit An orbit above about 700 miles (1126 km). At this altitude air drag is minimal.
Hydrogen bomb A bomb which derives its power from joining together atoms under extreme temperatures and pressures. The first test device used liquid hydrogen; present weapons use tritium, lithium hydrides or beryllium. An atomic bomb is used to generate the temperatures needed to start the process.

ICBM (Intercontinental Ballistic Missile) A rocket with a range of several thousand miles.
Infrared Electromagnetic radiation of wavelengths from the red end of the visible color spectrum to the microwaves used in radar.
Interstage fairing The part of a rocket which allows two stages to be joined together.
IRBM (Intermediate Range Ballistic Missile) A rocket with a range of about 1500 nautical miles (2413 km).

Kilojoules 1000 watts per second.
Kiloton The explosive effect of 1000 tons of TNT.

MAD or Mutual Assured Destruction A doctrine which holds that no matter how either superpower might attack the other, enough nuclear weapons will remain to destroy the aggressor in retaliation. This doctrine also insists that a nuclear war cannot be fought, won or even survived.
Megaton The explosive effect of one million tons of TNT.
Memory Component of a computer or guidance system which records and stores instructions and data.
Micrometeoroid Meteoroids less than 1/250th of an inch in diameter.
Microradian 0.0000573 degrees.
Miss distance The distance separating a satellite target and interceptor at the point of closest approach.
MX Missile A new US 3-stage solid-fuel ICBM carrying 10 warheads, also called 'Peacekeeper'.

Nautical mile 6060 ft – 1.15 statute miles or 1.852 km.
Newtonian telescope A type of reflecting telescope developed by Sir Isaac Newton. The light is reflected from a concave main mirror onto a small flat mirror held at a 45° angle and then out of the side of the telescope. In non-Newtonian telescopes the flat mirror is replaced by a small convex mirror that reflects the light back down the telescope and through a hole in the main mirror.

Offset oscillator frequency In Navstar, a slight, intentional adjustment to the transmitted signals to compensate for the effects of gravity.
Oxidizer In a rocket propellant, a substance such as liquid oxygen or nitric acid that yields oxygen for burning the fuel.

Payload An object (warhead or satellite) carried by a rocket.
Perigee The low point of a satellite's orbit. The point at which it is closest to Earth.
Proton The proton is an elementary particle and a constituent of every atomic nucleus. The hydrogen nucleus is in fact a single proton; but all other nuclei contain at least one proton and one neutron.

Radar cross section A measure of the radar echo which any particular object has.
Radio frequencies Normally expressed in kilohertz per second at and below 30 000 kHz, and megahertz per second above this frequency. Frequency subdivisions are: very low frequency

(VLF), below 30 kHz; low frequency (LF), 30 to 300 kHz; medium frequency (MF), 300 to 3000 kHz; high frequency (HF), 3000 to 30 000 kHz; very high frequency (VHF), 30 to 300 MHz; ultra-high frequency (UHF), 300 to 3000 MHz; super-high frequency (SHF), 3000 to 30 000 MHz; extremely high frequency (EHF), above 30 000 MHz. During World War II, radio-frequency bands were designated by letters (eg, K band, L band, P band, Q band, S band, V band, and X band). These designations were used originally to maintain military secrecy but currently have no official standing.

Real time Live transmission from a satellite to the ground. For there to be real time communications via satellite between two points on the ground the satellite must be above the horizon of both points. This is why communications satellites use higher orbits than other types.

Re-entry The return of a spacecraft which enters the atmosphere after flight above it.

Relativity According to Einstein's Theory of Relativity, time slows down increasingly in systems (eg, extremely high-performance spaceships) moving at velocities approaching the speed of light, relative to other systems in space (eg, the Earth).

Retrorocket A rocket that gives thrust in a direction opposite to the direction of the object's motion.

Satellite photo resolution A measure of the smallest object which can be seen on a photograph taken by an observation satellite. At the present time the best resolution is under 6 in (15 cm).

S band A radio frequency band of 1550 to 5200 megahertz per second.

Shroud A covering or nose cone that protects a satellite from aerodynamic pressure and heating as the rocket climbs through the atmosphere.

Solar cell A device which converts sunlight into electrical power.

Solid propellant Rocket propellant containing fuel and other oxidizer combined into a solid plastic-like cake called a *grain*.

Station-keeping Remaining in a precise orbit with a constant velocity.

Storable propellant A propellant in which fuel and oxidant can be stored in a rocket awaiting launch for long periods. Eg, *fuel:* a blend of hydrazine and unsymmetrical dimethyl-hydrazine (UDMH); *oxidant:* nitrogen tetroxide, which ignite on contact.

Subsystem One part of a larger system. A gyro-stabiliser is a sub-system of the guidance and control system.

Sun-synchronous A near-polar orbit which has its plane directed at the Sun.

Telemetry Radio signals from a rocket or satellite, received by a ground station, which indicate the function of the on-board systems.

Terminal guidance Guidance of the last phase of spacecraft rendezvous.

Thermal signature The particular heat characteristics of a satellite. Radiators will be at a high temperature while the fuel tanks tend to be cold.

Thermonuclear A term relating to the hydrogen bomb and the very high temperatures which it relies on to start the nuclear reaction.

Time ranging Determining the distance of a satellite from a ground-based tracking station.

Transistor An electronic device that controls an electron current by the conducting properties of germanium or like material.

Verniers Small thrusters used to adjust the flight path of a rocket or satellite.

Source Notes

Introduction

1. *Soviet Military Power* Department of Defense (DoD). 2. Thomas P Stafford, Lt-Gen (ret). 3. For background to this remark, see *Space Law 1957-67*. 4. *Pravda* mid-1982. 5. SALT 1 Treaty, 1972. 6. Ronald Reagan, tele-vised speech on US military policy, March 23, 1983. 7. *The Illustrated Encyclopedia of Space Technology* K W Gatland, Salamander, 1981, page 214. 8. *The Illustrated Encyclopedia of Space Technology* op cit, page 214. 9. 'A Soviet Space Shuttle?', *Spaceflight* K W Gatland, Sept-Oct 1978, page 325.

Chapter 1 Military Space Systems

Chapter opening quote from *On War* Carl Von Clausewitz, Penguin edition, 1968, page 118.

1. *Soviet Space Programs 1971-1975 Vol 1* US Library of Congress, 1976, page 392. 2. *Aviation Week & Space Technology* McGraw Hill, Inc., New York, October 6, 1980. 3. *Soviet Space Programs 1971-1975* op cit, Chapter 6 and Chapter 6, Annex 2. 4. *Aviation Week & Space Technology* op cit, September 29, 1980. 5. *Red Star in Orbit* James Oberg, Random House, 1981, Chapter 8. 6. *Spaceflight* The British Interplanetary Society, October 1981. 7. *Handbook of Soviet Manned Spaceflight* Nicholas L Johnson, American Astronautical Society, 1980, Chapter 12. 8. *Aviation Week & Space Technology* op cit, February 25, 1974. 9. *Secret Sentries in Space* Philip J Klass, Random House, 1971, Chapter 19. *US Intelligence and the Soviet Strategic Threat* Laurence Freedman, Westview Press, 1977, Chapter 4. 10. *The Falcon and The Snowman* Robert Lindsay, Pocket Books, 1980, page 127. 11. *Soviet Space Programs 1971-1975* op cit, page 406. 12. *Soviet Space Programs 1971-1975* op cit, page 434. 13. *Soviet Space Programs 1971-1975* op cit, pages 430-433. 14-15 *Spaceflight* op cit, May 1981. 16. *Aviation Week & Space Technology* op cit, September 10, 1973. 17. *Aviation Week & Space Technology* op cit, June 28, 1976. 18. *Spaceflight* op cit, July 1978. *Aviation Week & Space Technology* op cit, October 6, 1980. *Aviation Week & Space Technology* op cit, April 16, 1979. 19. *Journal of the British Inter-planetary Society* January 1982. 20. *Combat Missilemen* James Barr and William Howard, Harcourt, Brace and World, 1961, pages 24-25. 21. *Secret Sentries in Space* op cit, pages 173-177, *Journal of the British Interplanetary Society* January 1982. 22. *Aviation Week & Space Technology* op cit, December 2, 1974. 23. *Aviation Week & Space Technology* op cit, May 4, 1981. 24. *Aviation Week & Space Technology* op cit, February 28, 1978. 25. *Soviet Space Programs 1971-1975* op cit, pages 436-437. 26. *Secret Sentries in Space* op cit, page 183. 27. *The Raid* Benjamin F Schemmer, Harper and Row, 1976, page 137. 28. *Secret Sentries in Space* op cit, pages 140-141. 29. *The Aerospace Corporation: Its Work 1960-1980* Times Mirror Press, 1980, page 64. 30. *Spaceflight* op cit, February 1979. 31. *Soviet Space Programs 1971-1975* op cit, pages 366, 406. 32. *Space-flight* op cit, February 1982. 33. *Spaceflight* op cit, January 1980. 34. *Aviation Week & Space*

Technology op cit, November 8, 1982. 35. *The Aerospace Corporation: Its Work 1960-1980* op cit, Chapter 6. 36. *Jane's Pocket Book of Space Exploration* T M Wilding-White, Collier Books, 1976, page 182. 37. *Spaceflight Satellite Digest 101* February 1977. 38. *Soviet Space Programs 1971-1975* op cit, pages 345-351, 381. 39. *Flight International* January 2, 1982. 40. *Soviet Space Programs 1971-1975* op cit, page 408. 41. *Journal of the British Interplanetary Society* July 1981. 42. *The Soviet War Machine* Salamander Books, 1976, page 40. 43. *Soviet Space Programs 1971-1975* op cit, pages 386, 407. 44. *Newsweek* September 8, 1975. 45. *Reader's Digest* March 1982.

Chapter 2 The Nuclear Factor

Chapter opening quote from *The Fate of the Earth,* Jonathan Schell, Avon Books, 1982, page 188.

1. *Rockets, Missiles and Men in Space* Willy Ley, Signet, 1969. 2. *Space Travel* K Gatland and A Kunesch, Philosophical Library, 1953, page 102. 3. *Aviation Week & Space Technology* op cit, September 18, 1961. 4. *Secret Sentries in Space,* op cit, pages 11-15. 5. *A History of the US Air Force Ballistic Missiles* Ernest G Schwiebert, Praeger, 1965, Chapter 4. 6. *Across the Space Frontier* Cornelius Ryan, Editor, Viking Press, 1952, Chapter 2. 7. *Space Travel* op cit, pages 106-115. 8. *Astronautics in the Sixties* Ken Gatland, John Wiley & Sons, Inc, 1962, pages 146, 154. 9. *Soviet Space Programs 1971-1975* op cit, pages 400-402. 10. *Aviation Week & Space Technology* op cit December 5, 1960. 11. *Science News* November 18, 1967. 12. *Space Weapons: A Handbook of Military Astronautics* The Editors of Air Force Magazine, Praeger, 1959, pages 66-67. 13. *Spaceflight* op cit, April 1978. 14. *Spaceflight* op cit, November-December 1980. 15. *Aviation Week* op cit, September 16, 1957. 16. *Aviation Week & Space Technology* op cit, September 28, 1959. 17 *War for the Moon* Martin Caidin, Dutton, 1959, pages 113-115. 18.*Dyna-Soar Project Streamline* Carrollton Press, DOD 186A, 1977. 19. *Aviation Week & Space Technology* op cit, January 29, 1962. 20. *Aviation Week & Space Technology* op cit, April 10, 1961. 21. *Aviation Week & Space Technology* op cit, December 5, 1960. 22. *Aviation Week & Space Technology* op cit, December 12, 1960. 23. *Aviation Week & Space Technology* op cit, December 19, 1960. 24. *Missiles and Rockets,* March 26, 1962. 25. *Aviation Week & Space Technology* op cit, July 11, 1960. 26. *Aviation Week & Space Technology* op cit, February 6, 1961. 27. *New York Times* May 13, 1962, page 59. 28. *New York Times* September 6, 1962, page 16. 29. *US News and World Report* October 9, 1961. 30. *Time* August 23, 1963. 31. *Kennedy, Khrushchev and the Test Ban* Glenn T Seaborg, University of California Press, 1981, page 114. 32. *Secret Sentries in Space* op cit, pages 62-65. 33. *Saturday Evening Post* July 28, 1962. 34. *Newsweek* October 8, 1962. 35. *Spaceflight* op cit, December 1973 & March 1974. 36. *Missiles and Rockets* Ken Gatland, Macmillan, 1975, pages 189-192. 37. *Spaceflight* op cit, December 1976, Correspondence. 38. *Spaceflight* op cit, July/August 1977, Correspondence. 39. *Aviation Week & Space Technology* op cit, April 25, 1960. 40. *New York Times* February 11, 1960, page 1.

41. *New York Times* February 12, 1960, page 3. 42. *New York Times* February 24, 1960, page 14. 43. *New York Times* February 15, 1960, page 10. 44. *Aviation Week & Space Technology* op cit, April 2, 1962. 45. *Aviation Week & Space Technology* op cit, April 16, 1962. 46. *Aviation Week & Space Technology* op cit, October 15, 1962. 47. *Secret Sentries in Space* op cit, pages 22, 52. 48. *Lightning Bugs and Other Recon Drones,* William Wagner, Armed Forces Journal/Aero Publishers, 1982, page 19. 49. *Secret Sentries in Space* op cit, page 71. 50. *The Russians,* Hedrick Smith, Quadrangle, 1976, page 441. 51. *Khrushchev,* Mark Frankland, Stein and Day, 1967, page 187. 52. *Soviet Space Programs 1962-1965,* op cit, page 55. 53. *Khrushchev: A Career,* Edward Crankshaw, The Viking Press, 1966, page 270. 54. *Journal of the British Interplanetary Society,* August 1980. 55. *The Soviet War Machine,* Salamander Books, 1976, pages 205-206. 56. *Secret Sentries in Space,* op cit, page 109. 57. *Saturday Evening Post,* July 28, 1962. 58. *War and Space,* Robert Salkeld, Prentice-Hall, 1970, pages 125-126. 59. *Astronautics and Aeronautics 1963,* NASA SP-4004, page 65. 60. *Khrushchev: A Career,* op cit, pages 254-255, 263. 61. *Khrushchev,* op cit, page 84. 62. *Khrushchev,* op cit, pages 192-194. 63. *Time* August 23, 1963. 64. *Khrushchev,* op cit, page 208. 65. *Khrushchev,* op cit, pages 116, 166-168. 66. *Secret Sentries in Space,* op cit, pages 106-107. 67. *Carrollton Press* CIA 56A, 1975. 68. *Khrushchev,* op cit, pages 192-194. 69. *Khrushchev,* op cit, page 195. 70. *The Fall of Khrushchev,* Hyland and Shryock, Funk & Wagnalls, 1968, page 37. 71. *Khrushchev,* op cit, pages 281-283. 72. *The Brink,* David Detzer, Crowell, 1979, pages 48-54. 73. *Rockets & Missiles,* Bill Gunston, Salamander Books, 1979, page 53. 74. *Aerospace Technology* November 20, 1967. 75. *National Review* July 24, 1981. 76. *War and Space,* op cit, pages 152-153. 77. *The Soviet Estimate,* John Prados, The Dial Press, 1982, pages 204-205. 78. *The Soviet War Machine,* op cit, page 217. 79. *Soviet Space Programs 1971-1975,* op cit, page 418. 80. *Astronautics and Aeronautics 1965,* op cit, pages 521-522, 544. 81. *Soviet Space Programs 1971-1975,* op cit, page 378. 82. *Soviet Space Programs 1971-1975,* op cit, pages 418-422. 83. *Soviet Space Programs 1971-1975,* op cit, pages 416-417. 84. *Aerospace Technology* November 20, 1967. 85. *Spaceflight,* op cit, May 1981. 86. *Aviation Week & Space Technology,* op cit, October 16, 1967. 87. *Aviation Week & Space Technology,* op cit, December 11, 1967. 88. *Time* November 10, 1967. 89. *Astronautics and Aeronautics,* op cit, 1967, page 348. 90. *The Illustrated Encyclopedia of Space Technology,* Ken Gatland, Editor, Salamander Books, 1981, pages 40-41. 91. *The Soviet War Machine,* op cit, page 212. 92. *Soviet Space Programs 1971-1975,* op cit, pages 417, 420. 93. *Rockets & Missiles,* op cit, pages 172-173. 94. *Survival* September/October 1979. 95. *Weapons of World War 3,* William J Koenig, Crescent, 1981, page 99.

Sources for Space Law Chart (1957-1967) in Chronological order.

Secret Sentries in Space op cit, page 18. *War and Space* op cit, pages 135-136. *War and Space* op cit, pages 122-123. *New York Times,* March 17, 1960, page 1. *New York Times,* March 22, 1960, page 1. *New York Times,* April 5, 1960, page 1. *New York Times,* April 24, 1960, Section IV, page 9. *New York Times,* June 24, 1960, page 1. *New York Times,* September 26, 1961, pages 14-15. *New York Times,* October 14, 1961, page 6. *New York Times,* March 20, 1962, page 21. *New York Times,* June 8, 1962, page 4. *New York Times,* April 17, 1963, pages 1, 8. *New York Times,* May 3, 1963, page 3. *New York Times,* May 4, 1963, page 4. *New York Times,* June 19, 1963, page 3. *New York Times,* June 23, 1963, pages 1, 11. *New York Times,* August 15, 1963, page 12. *New York Times,* September 7, 1963, page 3. *New York Times,* September 10, 1963, page 11. *Secret Sentries in Space* op cit, page 127. *New York Times,* September 20, 1963, page 14. *New York Times,* October 2, 1963, page 14. *New York Times,* October 3, 1963, page 1. *New York Times,* October 4, 1963, page 1. *New York Times,* October 10, 1963, pages 16, 18. *New York Times,* October 16, 1963, pages 1, 3. *New York Times,* October 17, 1963, page 14. *Astronautics and Aeronautics 1963,* page 391. *New York Times,* October 18, 1963, page 30. *New York Times,* November 22, 1963, pages 18, 19. *Astronautics and Aeronautics 1963,* pages 457, 464, 477. *Astronautics and Aeronautics 1965,* NASA SP 4006, page 559. *Astronautics and Aeronautics 1966,* NASA SP 4007, pages 169, 198, 216, 237, 291. *Newsweek,* December 19, 1966. *Astronautics and Aeronautics 1966* op cit, pages 368-369, 377. *Astronautics and Aeronautics 1967,* NASA SP 4008, pages 23 and 126. *Space Law-Selected Basic Documents,* Library of Congress, 1976.

Chapter 3 The Origins of Anti-satellites

For chapter opening quote see source note 53.

1. *Carrollton Press,* DOD 39A, 1981. 2. *Missiles and Rockets* May 11, 1964. 3. *Missiles and Rockets* October 28, 1963. 4. *Spaceflight,* op cit, November 1977. 5. *Sky Rangers, Satellite Tracking Around the World,* Engle Drummond, John Day, 1965, Chapter 10. 6. *SPADATS Support of Program 437 Memorandum of Understanding.* 7. *Space Track, Watchdog of the Skies,* Charles Coombs, Morrow, 1969, pages 95-97. 8. *Sky and Telescope* February 1980. 9. *Sky Rangers,* op cit, page 151. 10. *Aerospace Technology* February 26, 1968. 11. *Aviation Week & Space Technology,* op cit, May 15, 1967. 12. *Aviation Week & Space Technology,* op cit, May 7, 1962. 13. *Weapons of World War 3,* op cit, page 52. 14. *Rockets & Missiles,* op cit, pages 138-139. 15. *Missiles and Rockets* February 4, 1963. 16. *Aviation Week & Space Technology,* op cit, October 8, 1969. 17. *New York Times* April 7, 1961, page 7. 18. *Aviation Week & Space Technology,* op cit, July 17, 1961. 19. *Aviation Week & Space Technology,* op cit, September 28, 1964. 20. *Missiles and Rockets,* May 11, 1964. 21. *Aviation Week & Space Technology,* op cit, March 20, 1961. 22. *Missiles and Rockets* September 28, 1964. 23. *Rockets & Missiles,* op cit, pages 170-172. 24. *ABM Research and Development at Bell Laboratories Project History* Chapters 1 and 2. 25. *Transcript of Secretary McNamara's Press Conference* September 18, 1964. 26. *Memorandum: Director of Defense Research and*

Engineering to Secretary of the Army, etc. Subject: Anti-Satellite Programs February 7, 1963. 27. *ABM Research and Development at Bell Laboratories* page 1-31. 28. *Minutes of Meeting-Briefing for Secretary McNamara on Satellite Detection, Inspection and Negation* June 27, 1963. 29. *ABM Research and Development at Bell Laboratories* page 1-32. 30. *DOD Weekly Summary* 14 April, 1964, Carrollton Press, DOD, 139D, 1979. 31. *Joint Chiefs of Staff message 241439Z* May, 1966. 32. *USAF Advanced Development Objective for an Anti-Satellite Program ADO-40.* 33. *Letter from Major General Momyer to ADC* May 31, 1962. 34. *The Mighty Thor,* Julian Hartt, Duell, Sloan and Pearce, 1961. 35. *Rockets & Missiles,* op cit, pages 61-62. 36. *New York Times* January 5, 1962, page 12. 37. *Aviation Week & Space Technology,* op cit, November 5, 1962. 38. *Aviation Week & Space Technology,* op cit, April 15, 1963. 39. *History of the Air Defense Command* July-December, 1964 Vol 1, pages 32-38. 40. *USAF to AFSC Early Satellite Interception Capability: Program 437* February 15, 1963. 41. *Memorandum for the Chief of Staff. Subject: Project 437* March 20, 1963. 42. *Memorandum for Deputy Chief of Staff Research & Development. Subject: Program 437* March 28, 1963. 43. *Letter from ADC to AF: Designation of User for Program 437* July 1, 1963. 44. *History of the Air Defense Command* July-December 1964, pages 45-46, 58-59. 45. *DOD Weekly Summary April 20, 1965.* Carrollton Press, DOD 145D, 1979. 46. *History of the Air Defense Command* July-December 1964, pages 55-57. 47. *History of the Air Defense Command* July-December 1964, pages 45, 49. 48. *DOD Weekly Summary, January 21, 1964,* Carrollton Press, DOD 138C, 1979. 49. *DOD Weekly Summary, February 18, 1964,* Carrollton Press, DOD 138D, 1979. 50. *DOD Weekly Summary, April 23, 1964,* Carrollton Press, DOD 139D, 1979. 51. *History of the Air Defense Command* July-December 1964, page 59. 52. *Astronautics and Aeronautics,* NASA SP 4005, page 320. 53. *History of the Air Defense Command* July-December 1964, page 60. 54. *Memorandum. Subject: Additional Boosters for Program 437,* September 15, 1965. 55. *History of the Air Defense Command* January-June 1966, page 296. 56. *Blue Suit Operation of Ground Guidance Station 6 and 7 at Vandenberg AFB* September 11, 1968. 57. *DOD Weekly Summary, April 4, 1967,* Carrollton Press, DOD 258D. 58. *DOD Weekly Summary April 11, 1967,* Carrollton Press, DOD 258E. 59. *Air Defense Command Assumes Responsibility for Thor Missile Launch Support of Space Programs* April 10, 1967. 60. *History of the Air Defense Command* July-December 1967, pages 302-303. 61. *Program Change Decision Z-7-085 Program 437,* December 28, 1967. 62. *History of the Air Defense Command* July-December 1967, pages 301-302. 63. *Message ADC to CSAF. Subject: Program 437 Combat Evaluation Launch* May 15, 1968. 64. *History of the Aerospace Defense Command FY69,* page 453. 65. *Program Management Directive for Program 437, Thor Missile Launch Support,* August 10, 1974. 66. *Aviation Week & Space Technology,* op cit, October 15, 1962. 67. *Aviation Week & Space Technology,* op cit, March 25, 1963. 68. *Appointment on the Moon,* Richard S Lewis, Ballantine Books 1969, pages 81, 308. 69. *Modifications to Existing Anti-Satellite Program and Follow On Programs,* August 6, 1963. 70.

Memorandum for the Chief of Staff. Subject: Satellite Interception, December 9, 1963. 71. *Memorandum: Satellite Interception,* December 20, 1963. 72. *Missiles and Rockets,* May 30, 1966. 73. *Aviation Week & Space Technology,* op cit, May 1, 1972. 74. *Letter from General McConnell USAF Chief of Staff to General Reeves ADC,* January 31, 1968. 75. *Memorandum of Understanding Special Defense Program,* November 1, 1968. 76. *History of the Aerospace Defense Command FY70,* page 343. 77. *Program 437 Combat Evaluation Launch,* March 1970. 78. *History of the Aerospace Defense Command FY70,* pages 353-355, *Memorandum. Subject: Johnston Island,* January 9, 1970, *Message. Subject: Program 437 Phasedown, Program 437 Concept of Operation,* January 21, 1970. 79. *ADC Programmed Action Directive 70-14,* November 1970, *Final Report Program 437. Damage Assessment Hurricane Celeste, Letter. Subject: GGS-2 (Johnston Island) Program 437,* December 15, 1972, *Message. Subject: Operational Status of GGS-2,* March 1973. 80. *Program 437 Cost Reduction Proposal,* November 2, 1973. 81. *Program Management Directive for Program 437, Thor Missile Launch Support,* August 10, 1974, *Program 437 Cost Reduction Proposal,* August 28, 1973, *Message CINCONAD to JCS,* March 6, 1975.

Chapter 4 The Anti-satellite Race

Chapter opening quote from *On War,* op cit, page 101.

1. *Secret Sentries in Space,* op cit, Chapter 10. 2. *Newsweek* November 20, 1960. 3. *Soviet Space Programs 1971-1975,* op cit, pages 105, 378-379. 4. *Secret Sentries in Space,* op cit, pages 109-110. 5. *Soviet Space Programs 1971-1975,* op cit, page 104. 5. *Soviet Space Programs 1971-1975,* op cit, page 104. 6. *Survival* January-February 1977 and *Soviet Military Strategy,* Marshal V D Sokolovsky, Prentice Hall, 1963. 7. *The Soviet War Machine,* op cit, Chapter 5. 8. *Soviet Military Power* DOD, 1981, pages 64-68. 9. *Aviation Week & Space Technology,* op cit, November 9, 1970. 10. *Newsweek* November 28, 1960. 11. *Rockets & Missiles* op cit, page 163. 12. *Foreign Nuclear Detections Through December 31, 1981,* Department of Energy, 1982. 13. *Space Radiation* William R Corliss, Department of Energy, 1968, page 44. 14. *The World in Space* United Nations, Prentice Hall, 1982. 15. *New York Times,* September 25, 1960, page 23. 16. *Missiles and Rockets,* November 14, 1960. 17. *Aviation Week & Space Technology,* op cit, December 10, 1962. 18. *Aviation Week & Space Technology,* op cit, November 14, 1960. 19. *Aviation Week & Space Technology,* op cit, December 5, 1960. 20. *New York Times,* November 20, 1960, page 39. 21. *Aviation Week & Space Technology,* op cit, August 1, 1960. 22. *Aviation Week & Space Technology,* op cit, June 26, 1961. 23. *Carrollton Press* NASA, 194A, 1977. 24. *Aviation Week & Space Technology,* op cit, December 10, 1962. 25. *Aviation Week & Space Technology,* op cit, December 17, 1962. 26. *Secret Sentries in Space,* op cit, page 124. 27. *Red Star in Orbit,* op cit, page 63. 28. *Survival* January-February 1977. 29. *Spaceflight,* op cit, September/October 1980. 30. *Aviation Week & Space Technology,* op cit, November 11, 1974. 31.

Soviet Space Programs 1971-1975 Vol II pages 188-192. 32. *History of the Aerospace Defense Command* FY1972, page 145. 33. *New Initiative (Project SPIKE)* September 14, 1971. 34. *Memorandum for Mr Hansen. Subject: Project SPIKE* September 21, 1971. 35. *Letter from General McKee to Commander ADC* July 9, 1971. 36. *Memorandum. Subject: SPIKE briefing at Headquarters USAF* September 21, 1971. 37. *Aviation Week & Space Technology,* op cit, April 26, 1976. 38. *Aviation Week & Space Technology,* op cit, June 21, 1976. 39. *Aviation Week & Space Technology,* March 28, 1977. 40. *Aviation Week & Space Technology* October 10, 1977. 41. *Aviation Week & Space Technology,* op cit, July 4, 1977. 42. *Survival* January-February 1977. 43. *War in Space,* James Canan, Harper & Row, 1982, page 21. 44. *New York Times* March 11, 1977, page 26. 45. *Aviation Week & Space Technology,* op cit, March 14, 1977. 46. *Aviation Week & Space Technology,* op cit, November 7, 1977. 47. *Aviation Week & Space Technology,* op cit, November 28, 1977. 48. *New York Times* March 19, 1979, page 1. 49. *Aviation Week & Space Technology,* op cit, March 27, 1978. 50. *New York Times* April 1, 1978, page 5. 51. *Spaceflight Satellite Digest,* October 1978. 52. *Aviation Week & Space Technology* op cit, April 17, 1978. 53. *War in Space,* op cit, pages 24-25. 54. *Aviation Week & Space Technology,* op cit, July 26, 1978. 55. *New York Times* June 18, 1978, page 8. 56. *New York Times,* February 20, 1979, Section IV, page 16. 57. *New York Times,* April 10, 1979, page 1. 58. *New York Times* April 23, 1979, page 5. 59. *New York Times,* June 1, 1979, page 6. 60. *Aviation Week & Space Technology,* op cit, July 9, 1979. 61. *The Russians,* op cit, page 264. 62. *Aviation Week & Space Technology* op cit, November 9, 1981. 63. *Aviation Week & Space Technology,* op cit, June 16, 1980. 64. *Aviation Week & Space Technology* op cit, January 18, 1982. 65. *Aviation Week & Space Technology,* op cit, May 3, 1982. 66. *Aviation Week & Space Technology,* op cit, February 8, 1982. 67. *Spaceflight,* op cit, September/October 1980. 68. *Aviation Week & Space Technology,* op cit, February 9, 1981. 69. *Spaceflight Satellite Digest 147,* August/September 1981. 70. *Spaceflight,* op cit, October 1981. 71. *Aviation Week & Space Technology,* op cit, March 23, 1981. 72. *Spaceflight Satellite Digest 152,* March 1982. 73. *Aviation Week & Space Technology,* op cit, October 26, 1981. 74. *Spaceflight* op cit, April 1982. 75. *Spaceflight* op cit, November 1982. 76. *Aviation Week & Space Technology* op cit, November 2, 1981. 77. *Aviation Week & Space Technology* op cit, September 27, 1982. 78. *Aviation Week & Space Technology* op cit, June 26, 1982. 79. *National Review,* July 24, 1981. 80. *War in Space* op cit, page 177. 81. *Aviation Week & Space Technology* op cit, June 28, 1982. 82. *Los Angeles Times,* July 5, 1982, page 1. 83. *Desert Wings,* Special Edition, Open House, 1982.

Chapter 5 Lasers

Chapter opening quote from *Illustrissimi,* Albino Luciani, Fount Paperbacks, 1979.

1. *The War of the Worlds,* H G Wells, Scholastic Book Services, 1968, pages 42-43. 2. *Day of Trinity,* Lansin Lamont, Atheneum, 1965, page 18. 3. *Lasers and Holography, Second Edition,* Winston E Kock, Dover, 1981,

pages 27-32. 4. *War in 2080,* David Langford, Morrow, 1979, page 67. 5. *Lasers and Holography* op cit, page 34. 6. *Secret Sentries in Space* op cit, pages 211-214. 7. *Aviation Week & Space Technology* op cit, August 18, 1975. 8. *Air Force Magazine,* August 1981. 9. *Aviation Week & Space Technology* op cit, August 4, 1980. 10. *Aviation Week & Space Technology* op cit, August 18, 1975. 11. *Aviation Week & Space Technology* op cit, August 4, 1980. 12. *Aviation Week & Space Technology* op cit, May 25, 1981. 13. *Aviation Week & Space Technology* op cit, September 8, 1975. 14. *Aviation Week & Space Technology* op cit, August 4, 1980. 15. *War in 2080* op cit, page 73. 16. *Aviation Week & Space Technology* op cit, August 18, 1975. 17. *Aviation Week & Space Technology* op cit, March 19, 1981. 18. *Aviation Week & Space Technology* op cit, August 4, 1980. 19. *Confrontation in Space,* G Harry Stine, Prentice Hall, 1981, page 117. 20. *Aviation Week & Space Technology* op cit, July 28, 1980. 21. *Spaceflight* op cit, May 1976. 22. *Robot Explorers,* Kenneth W Gatland, Macmillan, 1972, pages 136, 166, 172. 23. *Confrontation in Space* op cit, page 105. 24. *Daily Californian,* January 16, 1981, page B-1. 25. *Aviation Week & Space Technology* op cit, August 21, 1978. 26. *Aviation Week & Space Technology* op cit, September 8, 1975. 27. *Aviation Week & Space Technology* op cit, August 7, 1978. 28. *Air Force Magazine,* August 1981. 29. *Aviation Week & Space Technology* op cit, March 25, 1981. 30. *Aerospace Daily,* June 4, 1981. 31. *Aviation Week & Space Technology* op cit, September 8, 1975. 32. *Aviation Week & Space Technology* op cit, July 28, 1980. 33. *Aviation Week & Space Technology* op cit, February 16, 1981. 34. *Aviation Week & Space Technology* op cit, August 4, 1980. 35. *Aviation Week & Space Technology* op cit, August 21, 1978. 36. *Aviation Week & Space Technology* op cit, August 28, 1980. 37. *Confrontation in Space* op cit, Chapters 12, 13. 38. *Reader's Digest* March 1971. 39. *Aviation Week & Space Technology* op cit, September 8, 1975. 40. *War in 2080* op cit, pages 74-75. 41. *Aviation Week & Space Technology* op cit, February 16, 1981. 42. *Aviation Week & Space Technology* op cit, October 24, 1977. 43. *Aviation Week & Space Technology* op cit, November 22, 1982. 44. *Aviation Week & Space Technology* op cit, May 25, 1981. 45. *Air Force Magazine,* February 1982. 46. *ABM Research and Development at Bell Laboratories* op cit, page 1-37. 47. *Aviation Week & Space Technology* op cit, December 4, 1961. 48. *New York Times,* March 29, 1962, page 2. 49. *Missiles and Rockets* February 4, 1963. 50. *ABM Research and Development at Bell Laboratories* op cit, Chapter 1. 51. *Aviation Week & Space Technology* op cit, September 8, 1975. 52. *Aviation Week & Space Technology* op cit, February 16, 1981. 53. *Rockets & Missiles* op cit, pages 172-173. 54. *Atlas – The Story of a Missile,* John L Chapman, Harper & Brothers, 1960, Chapter 11. 55. *Aviation Week & Space Technology* op cit, July 28, 1980. 56. *Aviation Week & Space Technology* op cit, August 4, 1980. 57. *Confrontation in Space* op cit, page 105. 58. *Aviation Week & Space Technology* op cit, April 12, 1982. 59. *Soviet Military Power* op cit, pages 84-85. 60. *Aviation Week & Space Technology* op cit, April 12, 1982. 61. *Aviation Week & Space Technology* op cit, August 21, 1978. 62. *Aviation Week & Space Technology* op

cit, May 25, 1981. 63. *Aviation Week & Space Technology* op cit, July 28, 1980. 64. *Aviation Week & Space Technology* op cit, September 27, 1982. 65. *Spaceflight* op cit, September/October 1978. 66. *Aviation Week & Space Technology* op cit, May 25, 1981. 67. *Air Force Magazine* August 1981. 68. *Aviation Week & Space Technology* op cit, May 15, 1967. 69. *Aviation Week & Space Technology* op cit, July 28, 1980. 70. *Aviation Week & Space Technology* op cit, August 7, 1978. 71. *Aviation Week & Space Technology* op cit, August 20, 1978. 72. *Aviation Week & Space Technology* op cit, July 28, 1980. 73. *Aviation Week & Space Technology* op cit, April 12, 1982. 74. *Aviation Week & Space Technology* op cit, July 28, 1980. 75. *Aviation Week & Space Technology* op cit, July 28, 1980. 76. *Aviation Week & Space Technology* op cit, April 12, 1982. 77. *Aviation Week & Space Technology* op cit, May 25, 1981. 78. *Aviation Week & Space Technology* op cit, July 28, 1980. 79. *Soviet Space Programs 1971-1975* op cit, page 610. 80. *Aviation Week & Space Technology* op cit, June 14, 1982. 81. *Aviation Week & Space Technology* op cit, July 20, 1981. 81. *Aviation Week & Space Technology* op cit, May 25, 1981. 83. *Aviation Week & Space Technology* op cit, August 4, 1980. 84. *Aviation Week & Space Technology* op cit, February 16, 1981. 85. *Technology Review,* October 1981, *Air University Review,* March/April 1982, *Aviation Week & Space Technology* op cit, July 28, 1980, *Air Force Magazine,* August 1981. 86. *Aviation Week & Space Technology* op cit, April 12, 1982. 87. *Air Force Magazine,* August 1981. 88. *Science 83* January/February 1983. 89. *Working Paper. Subject: Anti-Satellite Systems,* May 4, 1970. 90. *San Diego Tribune,* November 26, 1982, page Y-6. 91. *San Diego Union,* October 31, 1982, page C-4. 92. *Aviation Week & Space Technology* op cit, April 12, 1982. 93. *Air Force Magazine,* August 1981. 94. *Aviation Week & Space Technology* op cit, July 28, 1980. 95. *Aviation Week & Space Technology* op cit, May 25, 1981. 96. *Aviation Week & Space Technology* op cit, April 12, 1982. 97. *Strategic Air Command,* David A Anderson, Scribers, 1975, page 138. 98. *Aviation Week & Space Technology* op cit, May 25, 1981. 99. *Aviation Week & Space Technology* op cit, April 12, 1982. 100. *Aviation Week & Space Technology* op cit, May 25, 1981. 101. *Space World,* August/September, 1982.

Chapter 6 Particle Beam Weapons

For chapter opening quote see source note 38.

1. *Aviation Week & Space Technology* op cit, August 4, 1980. 2. *Elementary Modern Physics,* Atam P Arya, Addison Wesley, 1974, page 445. 3. *War in 2080* op cit, page 81. 4. *Aviation Week & Space Technology* op cit, August 4, 1980. 5. *Flight International,* April 24, 1982. 6. *War in 2080* op cit, page 82. 7. *Aviation Week & Space Technology* op cit, May 25, 1981. 8. *Aviation Week & Space Technology* op cit, July 28, 1980. 9. *War in 2080* op cit, page 80. 10. *Flight International,* April 24, 1982. 11. *Aviation Week & Space Technology* op cit, July 28, 1980. 12. *Aviation Week & Space Technology* op cit, August 4, 1980. 13. *Elementary Modern Physics* op cit, page 435. 14. *Aviation Week & Space Technology* op cit, August 4, 1980. 15. *War in 2080* op cit, page 81. 16. *Confrontation in Space* op cit, pages 115-116.

17. *Aviation Week & Space Technology* op cit, May 2, 1977. 18. *Spaceflight,* November/December 1980. 19. *Aviation Week & Space Technology* op cit, May 2, 1977. 20. *Aviation Week & Space Technology* op cit, October 8, 1979. 21. *Tora, Tora, Tora,* Gordon W Pranger, McGraw-Hill 1963. 22. *Aviation Week & Space Technology* op cit, July 28, 1980. 23. *Confrontation in Space* op cit, pages 110-111. 24. *Aviation Week & Space Technology* op cit, February 16, 1981. 25. *Aviation Week & Space Technology* op cit, July 28, 1980. 26. *Aviation Week & Space Technology* op cit, July 28, 1980. 27. *Aviation Week & Space Technology* op cit, August 4, 1980. 28. *Aviation Week & Space Technology* op cit, July 28, 1980. 29. *Air Force Magazine,* July 1981. 30. *Aviation Week & Space Technology* op cit, July 28, 1980. 31. *Air Force Magazine,* July 1981. 32. *Aviation Week & Space Technology* op cit, July 28, 1980. 33. *Aviation Week & Space Technology* op cit, August 4, 1980. 34. *Aviation Week & Space Technology* op cit, July 28, 1980, *Aviation Week & Space Technology* op cit, February 23, 1981. 35. *Weapons of World War 3* op cit, page 28. 36. *Air University Review,* March/April 1982. 37. *War and Space* op cit, pages 164-165. 38. *Shall America Be Defended,* Lt Gen Daniel O Graham, Arlington House, 1979, pages 63-64, 70. 39. *Weapons of World War 3* op cit, page 88. 40. *Rockets & Missiles* op cit, page 113. 41. *The Rocket,* Dave Baker, Crown, 1978, Chapter 10. 42. *Ballistic Missile Defense,* Benson D Adams, American Elsevier Publishing Company, 1971, page 240. 43. *Shall America Be Defended* op cit, page 72. 44. *Air University Review,* March/April 1982. 45. *Aviation Week & Space Technology* op cit, January 17, 1983. 46. *Armed Forces Journal,* August 1981 and January 1982. 47. *Air Progress Pilot Reports* 1978. 48. *War in Space* op cit, pages 62 and 178. 49. See *Spaceflight* op cit, April 1980 and June 1980 for a previous example. 50. *War in Space,* op cit, pages 150-151. 51. *Sea Power,* June 1982. 52. *Ballistic Missile Defense,* Chapter 8. 53. *Comparative Strategy* Vol 3, 1981, Chapter 2. 54. *War in Space* op cit, page 159. 55. *KGB,* John Barron, Bantam Books, 1974, pages 4-5, 243. 56. *The Chinese War Machine,* Salamander Books, 1979, page 127. 57. *Weapons of World War 3* op cit, page 100. 58. *Illustrated Story of World War II,* Reader's Digest, 1969, pages 68-77. 59. *Blitzkieg,* Robert Wernick, Time-Life Books, 1977, page 55. 60. *Aviation Week & Space Technology* op cit, July 28, 1980. 61. *Air University Review,* March/April 1982. 62. *Aviation Week & Space Technology* op cit, March 26, 1981. 63. *Aviation Week & Space Technology* op cit, July 20, 1981. 64. *Air University Review,* March/April, 1982.

Index

Picture Credits
Australian Defense Department *(Aviation Week)*/MARS 18; CERN *(New Scientist)*/MARS 160, 165; Curtis Peebles 53, 97, 131; *Daily Mirror*/MARS 62; Department of Defense/MARS 9, 13, 69; Ford Aerospace/MARS 44; Kenneth Gatland 19, 32, 33, 36, 64, 69, 82, 84, 100, 109, 150; Lawrence Livermore Laboratory 157; Lockheed Missile and Space Company, USA/MARS 40, 97; MOD Crown copyright/MARS 41; NASA/MARS 111, 150; Novosti/MARS 63, 114; RCA Astro-electronics/MARS 39; US Airforce/MARS 42, 54, 55, 58, 68, 70, 72, 83, 88, 122, 123, 136; US Army 71; US Navy 89

Illustrations
Terry Hadler 34-35, 122-123, 158-159; Hussein Hussein 12, 27, 42, 53, 59, 67, 79, 84, 89, 100, 108, 126, 136, 150, 157, 169 and all silhouettes; Simon Roulston 9, 31, 86; Mike Trim 70-71

Special thanks to Charles P Vick for rocket silhouette reference